Global Justice and the Biodiversity Crisis

Global Justice and the Biodiversity Crisis

Conservation in a World of Inequality

Chris Armstrong

OXFORD
UNIVERSITY PRESS

Great Clarendon Street, Oxford, OX2 6DP,
United Kingdom

Oxford University Press is a department of the University of Oxford.
It furthers the University's objective of excellence in research, scholarship,
and education by publishing worldwide. Oxford is a registered trade mark of
Oxford University Press in the UK and in certain other countries

Published in the United States of America by Oxford University Press
198 Madison Avenue, New York, NY 10016, United States of America

British Library Cataloguing in Publication Data
Data available

Library of Congress Control Number: 2023951623

ISBN 9780198853596

DOI: 10.1093/9780191888090.001.0001

Printed and bound in the UK by
Clays Ltd, Elcograf S.p.A.

Acknowledgements

For comments on various chapters—and in some cases the whole manuscript—I thank Catriona McKinnon, Nicole Hassoun, Alex McLaughlin, Laura García-Portela, Christian Barry, Pablo Magaña Fernandez, Jamie Draper, Julia Mosquera, Margaret Moore, and Colin Rowe, as well as the anonymous reviewers for Oxford University Press. I also thank audiences at the University of Fribourg, Nova University Lisbon, Nuffield College Oxford, Wuhan University, Exeter University, and the University of Oslo, where I presented work in progress. As the book neared completion, I benefited enormously from two dedicated manuscript workshops. The first took place at UCLouvain in September 2022. I thank Axel Gosseries, Juan Olano Azpiroz, Charles-Hubert Born, Anna Wienhues, Pierre André, and Jens Jørund Tyssedal for participating and for making that event such a helpful and convivial one. The second took place in January 2023, at the Forschungskolleg Humanwissenschaften at Bad Homburg, in coordination with the Goethe University Frankfurt. My thanks go to Darrel Moellendorf, Petra Gümplová, Hanna Schueber, Julio Caceda, and Lukas Sparenborg for enabling a relaxed but intellectually fertile discussion. A version of Chapter 4 previously appeared in *Conservation Biology*, and I am grateful to be able to reproduce it here.

I dedicate this book to my mother, Caroline Caddick, for all the love and support she has given to me.

Contents

Introduction

Biodiversity in crisis

We live in a time of massive ecological destruction, for which human beings are the chief culprits. In the grand sweep of the earth's history, for one species to have undermined the planet's ecosystems as ours has done appears to be a unique event. Frankly, even for *homo sapiens* it is a remarkable development. For more than ninety-five per cent of our existence we were just one more species of Great Ape trying to eke out a living in a wild and challenging planet; in little more than the blink of an eye (in evolutionary terms), our kind has come to transform much of the living world.

Human life itself has also transformed. The dawn of the agricultural revolution twelve or thirteen thousand years ago unleashed new forms of inequality and hierarchy in human society (Sterelny 2021), even if hierarchy had its own venerable pre-history (Graeber and Wengrow 2021). It also kick-started a remarkable increase in the human population. Later developments, including the Industrial Revolution and the birth of the joint stock company, spurred the global spread of capitalism, a transnational trade in human beings, and waves of violent dispossession across the planet.

The impact of these developments on the non-human world was, if anything, even more dramatic. Agriculturalists domesticated a relatively small number of mammal and bird species, and selectively bred them in staggering numbers. Settler colonialists crossed the seas with new crops and livestock, which they used to replace diverse ecosystems with intensively farmed monocultures. Hacking their way through the tropics and the pampas, they encountered animals that were defenceless against their traps and guns. One of the lesser-noted—but nevertheless hugely important—results of these waves of conquest was a deluge of anthropogenic extinctions, with an estimated 130,000 species of invertebrate being driven to extinction since the early modern period (Vettese 2020). As early as 1800, during his famous journey through Latin America, the naturalist and polymath Alexander von Humboldt was disturbed by the immense damage an economy based on colonial control, slavery, and monocropping was doing to wider ecosystems, to

Global Justice and the Biodiversity Crisis. Chris Armstrong, Oxford University Press.

weather patterns, and to the livelihoods of indigenous people (Wulf 2015). Later, the stitching together of the seas into a single world ocean—courtesy of Columbus and Da Gama, as well as the nameless labourers who would eventually carve out the Panama and Suez canals—brought about a mingling of species that has fundamentally transformed marine ecosystems (Molnar et al. 2008).

It is from the second half of the twentieth century onwards, however, that the human impact on the living world has been most dramatic. What scientists call the Great Acceleration has seen rapid increases in human populations, in countries' gross domestic product (GDP), in energy and water use, and in the mass transportation of goods. At the same time it has brought about a step change in major 'earth systems trends' including the degradation of land, desertification, tropical deforestation, and ocean acidification (Steffen et al. 2015). The result has been a multifaceted planetary crisis, one of the chief elements of which is a rapid increase in the rate of biodiversity loss. One study has found that wild animal populations, for example, declined in numbers by an average of 69 per cent between 1970 and 2014 (WWF 2022). According to the comprehensive Dasgupta Review on the Economics of Biodiversity, meanwhile, populations of birds, mammals, reptiles, and amphibians plummeted by 70 per cent in a similar period (Dasgupta 2021: 375). These diverse wild animal populations have largely been replaced by much more homogenous human-reared animals and crops, as part of what Mark Rowlands has pithily described as a 'massive biomass reallocation program' (Rowlands 2021: 185). Measured in terms of the total quantity of carbon in our bodies, the vast bulk of the world's mammals, for instance, now fit into three blocs, each of roughly similar size: humans; human-reared cattle; and other domesticated animals. Together these three blocs represent fully 96 per cent of the world's extant mammalian biomass. All of the world's wild mammals, by contrast, make up the other 4 per cent (Bar-On, Phillips, and Milo 2018). For this shrinking remainder, life on the margins looks increasingly bleak. Whether some parts of the planet still deserve the name 'wilderness' is hotly debated (Cronon 1996), but it is undoubtedly the case that non-domesticated animals are increasingly forced to live in our proximity, and are experiencing stress, hunger, and difficulty in rearing young as a result.

For some commentators, these dramatic changes raise the spectre of an 'Anthropocene' world. The concept of the Anthropocene originally emerged within geology, where it was claimed that humanity's impact on the planet was already significant enough, by the mid-twentieth century, that it will likely show up in the geological record of the earth long into the future

(Crutzen 2006). But the term has now gained much wider currency as an indicator of humanity's fundamental transformation of the earth's biosphere. Rather than a gradual intensification of our environmental impacts, it is intended to pick out a qualitative shift in which human activities have comprehensively reordered or even replaced a number of natural processes (see Dryzek and Pickering 2019: 2–6). The utility of the concept of the Anthropocene, though, has been contested. For humans are not equal in their impacts on the biosphere, and some lifestyles are more sustainable than others. The rival idea of the 'capitalocene' usefully draws attention to the class basis of environmental destruction, and the hugely unequal impacts 'we' have on the planet (Moore 2017). The 'plantationocene' concept, meanwhile, picks out the connections between biodiversity destruction, colonialism, and racial capitalism (Haraway 2015).

If the loss of non-domesticated biomass captures one dimension of our impact on the living world, species extinctions capture another. The twentieth century witnessed the disappearance of, among others, the Tasmanian Wolf, Darwin's Galapagos Mouse, the Desert Bandicoot, the Guam Flying Fox, the Tarpan, the Golden Toad, the Grand Cayman Oriole, the Kandar fish, Passenger Pigeon, and Caribbean Monk Seal. Others are under serious pressure. Every species of Great Ape (apart from *homo sapiens*) is now considered endangered, with most qualifying as critically endangered (Kühl et al. 2017). The story of loss, however, is much broader. Media attention is often disproportionately focused on threats to 'charismatic' wildlife, including gorillas, pandas, elephants, and rhinos. But these, along with famous cases like the Tasmanian Wolf, are only the tip of a much greater extinction iceberg. In the twenty-first century, fully a quarter of all vertebrate species are threatened with extinction (Allan et al. 2019), and the ranks of the International Union for the Conservation of Nature's 'Red List' of threatened species grows longer with each update.[1] Many scientists now openly claim that we are entering a Sixth Great Extinction event, on a par with other cataclysmic events that have peppered the planet's past (Kolbert 2014).[2]

On dry land, the greatest *cause* of biodiversity loss has been habitat destruction (IPBES 2019: 12, Maxwell et al. 2010), especially in the interest of agriculture (Corlett 2020: 225). Scientists have estimated that while a mere 15 per cent of the earth's surface was used by humans a century ago, that figure now stands at 77 per cent (Watson et al. 2018a: 27). It may be more: according to another estimate, humanity now directly impacts 84 per cent

---

[1] https://www.iucnredlist.org/about/barometer-of-life (Accessed 29 November 2022).
[2] For critical discussion of some of the uncertainties around this claim, see Newman, Varner, and Linquist 2017: 7–13.

of the earth's land surface (Abegão 2019: 2). Measures of 'use' are of course contested, and in the colonial era indigenous peoples' distinctive ways of engaging with the land were wrongly depicted as having no normative consequence, before being all but written out of history. But it is still bracing to learn that over 85 per cent of the world's wetland, for instance, has been destroyed since 1970 (IPBES 2019: 11). At the same time, areas without significant human pressure are clearly declining rapidly (Venter et al. 2016: 5), not least given the truly global impact of climate change. For many of the species with which we share the earth, human expansion represents a truly catastrophic process. As the capitalist juggernaut rumbles on, 40 per cent of the planet's ice-free surface is now used for agriculture and livestock production (Machovina et al. 2015). Logging is also a major threat to biodiversity (Venter et al. 2016: 5), with as many as 80,000 acres of rainforest destroyed every day (Hale 2016: 211), much of it cleared to feed cattle for export. Since this land is often quickly exhausted of nutrients, the bulldozers push further onwards. If human demand for food doubles or even trebles by 2100, as some scientists have predicted (Clay 2011), the pressure on habitat will likely only intensify. Although some animals may find a new liminal existence on the edges of our societies (like the foxes that pick through litter bins at night, or the pigeons that throng public squares), most will find themselves outcompeted for food, fresh water, and living space.

Aside from habitat loss, other key drivers of biodiversity loss on land include climate change, pollution, and the spread of invasive species, each of which are putting serious pressure on many ecosystems (IPBES 2019). In the ocean, by contrast, fishing has had the greatest negative impact on biodiversity, followed by habitat destruction, climate change, pollution, and invasive species (IPBES 2019: 28–9). Industrial fishing has wrought a truly devastating impact on both the seabed and the water column, with populations of large fish, for instance, down in size by around 75 per cent over the last century (Christensen et al. 2014). Habitat destruction has seen widespread losses of key ecosystems such as kelp forests, coral reefs, and mangroves, each of which frequently act as crucial nurseries for marine life. But climate change is also wreaking havoc by acidifying the ocean's waters (and making it harder for calcifying organisms to grow or to recover from damage), by warming its waters (a major problem for the many marine organisms that can survive only within a narrow thermal envelope), and by intensifying extreme weather patterns. As these pressures continue, the human impact on the sea raises the spectre of an Anthropocene Ocean, dominated by large-scale fish farms and extractive industries, with its increasingly warm and acidic depths populated by a few resilient non-domesticated species like jellyfish and squid (Armstrong 2022: 223).

In this book, I will talk of a biodiversity *crisis*.[3] Scientists often diagnose (and seek to explain) empirical trends or shifts from one state of affairs to another. But to describe a trend as a crisis is to move from purely descriptive language to a normative register, and to point to *objectionable* or regrettable features of a given trend. As I employ it in this book, the idea of a biodiversity crisis rests on several key claims, which are, respectively, empirical, norma-tive, and political in character. First, the idea suggests that rates of biodiversity loss are accelerating very substantially beyond the expected or 'background' rate, on a number of dimensions (including loss of genetic diversity, loss of species, and loss of ecosystem variation). Second, it suggests that this accelerated loss of biodiversity threatens to have serious and objectionable consequences for human beings, and for members of other species. We will put more flesh on the bones of this claim in Chapter 2, where we explore the relationship between biodiversity loss and the survival, and flourishing, of humans and non-human animals. Third, the idea suggests that political responses to date have failed to make a substantial impact on the rate of bio-diversity loss, which is in fact accelerating despite those responses (more on this later). In that sense biodiversity loss represents both a moral crisis—in light of its very grave potential consequences—as well as a political one, in light of the failure of collective institutions thus far to rise to the challenge.

The biodiversity crisis as a distinct global challenge

In recent years the biodiversity crisis has achieved increasing prominence within politics and the mass media. In the autumn of 2021, for example, reports that the world had lost roughly half of its biodiversity gained consid-erable media coverage.[4] Nevertheless, the biodiversity crisis has often been the poor cousin of climate change when it comes to capturing popular imag-ination and concern, as well as political attention.[5] Both issues came to wide public prominence around the same time (roughly coinciding with the Rio Earth Summit of 1992), and they have lately been called 'the twin crises of the

---

[3] There are a large number of scientists who currently address this same problem. Most obviously, the notion of crisis is written deeply into the history of conservation biology as a discipline. See Soulé 1985 for a highly influential account of conservation biology as an inherently crisis-oriented, ethically driven endeavour.

[4] The occasion was the launch of the Biodiversity Intactness Index, prepared in conjunction with the UK's Natural History Museum. https://www.nhm.ac.uk/our-science/data/biodiversity-indicators/about-the-biodiversity-intactness-index.html (Accessed 12 October 2021).

[5] For a useful exploration of why this might have been the case, see Zaccai and Adams 2012.

Anthropocene' (Corlett 2020: 221). The year 1992 saw the birth of two key environmental Conventions: the UN Framework Convention on Climate Change (UNFCCC), and the Convention on Biological Diversity (CBD). But in the years since Rio, climate change has been considerably more successful in sustaining high levels of public interest (Youatt 2015: 6; see also Maxwell et al. 2010). According to a 2018 study, for example, the climate crisis receives eight times as much mainstream media coverage as biodiversity loss (Legagneux et al. 2018). National leaders are widely expected to attend the regular Conferences of the Parties on the climate; the same, however, cannot be said for the version focusing on biodiversity.[6] One especially gloomy pronouncement holds that the 1990s were 'the high point of interest, both scholarly and public, in biodiversity. There is no movement now' (Kotsakis 2021: x). Recent signs that attention to the biodiversity crisis is accelerating notwithstanding, there is still a very long way to go in comparison to climate change, both in terms of public attention and institutional responses.

The two challenges are of course bound up together—and indeed with other dimensions of a wider planetary crisis, including pollution, and the disruption of the nitrogen and phosphorus cycles (Steffen et al. 2015). For now, though, consider the connections between biodiversity and the climate. The destruction of biodiversity is both a cause of climate change (since the razing of carbon sinks like forests, peatlands, and mangroves spurs further warming), and a consequence of it (as higher temperatures, extremes of weather, and ocean acidification strain the ability of many plants and animals to adapt). But the crises are nevertheless at least partly separate, and it is therefore vital that each receives sufficient attention, rather than one being treated as a simple epiphenomenon of the other. If it is true that climate change has dominated the political and philosophical agenda to a considerable extent in recent years, crowding out attention to other environmental problems that might be just as serious (Noss et al. 2012), this would be highly regrettable. After all, the world's leaders could 'solve' the problem of climate change and still face a serious biodiversity crisis. Since it is land use change, rather than climate change, that represents the biggest threat to global biodiversity (Noss et al. 2012, Maxwell et al. 2010, Newbold et al. 2016, Rands et al. 2010), it is likely that even if policy makers stabilized the planet's climate, biodiversity would still be under serious threat. Although industrial agriculture is the biggest challenge, fishing, hunting, and pollution also pose threats to biodiversity that are (to a significant extent) independent of climate change.

---

[6] In 2022, the UK Prime Minister Rishi Sunak was widely criticized for his initial hesitation about attending climate COP27. But the fact that he did not attend biodiversity COP16, later the same year, attracted very little comment.

On the other hand, some proposed solutions to climate change would themselves threaten biodiversity. Consider the ongoing replacement of ages-old rainforest with monocultural tracts of trees selected for their rapid carbon-absorptive capacities. In South America these plantations have been called 'green deserts', because they are so much less hospitable to biodiversity than the rainforests they are steadily supplanting (Bremer and Farley 2010). As the Intergovernmental Panel on Biodiversity and Ecosystem Services (IPBES) cautiously puts it, 'the large-scale deployment of bioenergy plantations and afforestation of non-forest ecosystems can come with negative effects for biodiversity' (IPBES 2019: 18). Or consider the challenges that some geoengineering technologies such as ocean fertilization or atmospheric aerosol injection could pose to existing ecosystems by altering the chemical makeup of the ocean or the availability of light.[7] If political responses to climate change are to be pro-biodiversity, this will have to happen by design: we cannot simply assume that what assuages one problem must by definition alleviate the other.

Chapter 2 considers the dangers posed by the biodiversity crisis much more carefully. But given the role of biodiversity in supporting all of life on earth, responding effectively to its accelerating destruction may actually be the most important political challenge the world faces today. It is becoming ever clearer, however, that existing conservation policies are insufficiently aligned to the scale of the problem. The loss of species and of biodiverse habitat has not been halted (or even significantly slowed) despite decades of effort (Marvier et al. 2012, Leclere et al. 2020). There have certainly been some conservation success stories along the way. According to one study, conservation efforts since 1993 have prevented the extinction of 21–32 bird species, and 7–16 species of mammal (Bolam et al. 2021). But species loss nevertheless continues apace, at an estimated one thousand times the expected 'background' rate (De Vos et al. 2015), and genetic and ecosystem variation are diminishing rapidly. Business as usual is simply failing to get to grips with the scale of the crisis.

Beyond conservation business as usual

To date, conservation politics has tended to be rather piecemeal in nature. It has often focused, for example, on the preservation of particular species in

[7] Ocean fertilization involves adding nutrients to the top layer of the ocean, to stimulate phytoplankton growth. Atmospheric aerosol injection involves adding chemicals to the stratosphere, to increase its reflectiveness, or albedo.

one habitat or another. Timely action has resulted in many small victories—but they represent the exception rather than the rule, and push against a tide of accelerating biodiversity loss. In response, many policy makers have come to embrace more ambitious and synoptic goals—and conservation scholars and practitioners have urged them to aim still higher. The most important goal-setting framework to date has been the 1992 CBD. Under its auspices, the Aichi Biodiversity Framework aimed to meet twenty key targets during the period 2011 to 2020. One of them, for instance, was to protect 17 per cent of terrestrial areas, and 10 per cent of coastal and marine areas.[8] Not a single one of its twenty targets was fully met, however, and the majority were not even partly met. The Kunming–Montreal Global Biodiversity Framework[9] (adopted in December 2022) set its sights still higher, aiming to protect 30 per cent of terrestrial and marine ecosystems by 2030 (the '30 by 30' target, which has garnered much support in recent years[10]). Targets on this kind of scale, it is said, might help stabilize the loss of biodiversity by 2030, and perhaps even reverse it by 2050.[11]

Judging from existing national policies, the 30 by 30 target is also likely to be missed. But leading conservation scholars have in any case argued that a goal of 30 per cent ecosystem protection is not bold enough to stem the tide of biodiversity loss (see e.g. Watson et al. 2021, Allan et al. 2022). According to the IPBES, key biodiversity goals 'may only be achieved through transformative changes across economic, social, political and technological factors', and cannot be met via current trajectories (IPBES 2019: 14). The wider literature on conservation biology, political ecology, and environmental philosophy have all witnessed calls 'to raise ambition and forge a new transformative global plan for biodiversity' (Soto-Navarro et al. 2021: 935), and a focus on radical and transformative change has rapidly come to be central to these discussions (see e.g. Wyborn et al. 2020a, Massarella et al. 2021, Lundquist 2021, Leadley et al. 2022). But what would radical and transformative visions for biodiversity look like? Perhaps most famously, many academics and activists have argued for the protection of 50 per cent of the world's ecoregions, under the disparate banners of 'Half Earth', 'Nature Needs Half', or a 'Global Deal for Nature' (see e.g. Wilson 2016, Cafaro et al. 2017, Dinerstein et al. 2017,

---

[8] https://www.cbd.int/sp/targets/rationale/target-11/
[9] https://www.cbd.int/doc/c/e6d3/cd1d/daf663719a03902a9b116c34/cop-15-l-25-en.pdf
[10] Parallel national targets have also been adopted by a number of countries (including the US, which is, famously, not a party to the Convention on Biological Diversity. See https://www.nationalgeographic.com/environment/article/biden-commits-to-30-by-2030-conservation-executive-orders (Accessed 19 January 2022)).
[11] https://www.cbd.int/doc/c/abb5/591f/2e46096d3f0330b08ce87a45/wg2020-03-03-en.pdf (Accessed 16 January 2022).

Kopnina et al. 2018, Dinerstein et al. 2020). Because extant biodiversity is geographically concentrated, protecting 50 per cent of ecoregions might allow us to preserve up to 85 per cent of species, for instance, and thereby stave off the extinction crisis (Wilson 2016). Locke (2014), in fact, suggests that protecting 50 per cent of ecoregions is a mid-point estimate about what would be required to head off the biodiversity crisis, and that in reality the world might need to protect considerably more.[12]

I will not attempt to resolve the question of what precise proportion of ecosystems ought to be protected. One reason is that predicting the likely impact of any particular conservation measure would demand an advanced understanding of highly complex ecological processes. Even conservation biologists cannot be certain about the effects of protecting, say, 30 or 50 per cent of ecoregions. Wilhere (2021) has argued that an 'IPCC-like' scientific effort to pool relevant knowledge would be required before such proposals could be properly evaluated, and that building such knowledge ought to be a political priority. I do believe, however, that proposals on this scale of ambition are plausible, *if* political leaders decide that they want to secure a healthy and liveable ecosystem both for members of our species and for the other creatures we share the world with. Neither will I seek to resolve how much biodiversity ought to be conserved—whether 80 per cent, 100 per cent, or even more, as the tide of destruction is reversed. The point is that the general *prima facie* plausibility of these suggestions—and, crucially, the fact that they are already shaping global conservation politics—means that we urgently need to examine their implications for global justice. Even if it transpires that justice 'only' requires the conservation of 50 per cent of biodiversity, say, the questions I address in this book would still be of interest. If we should be still more ambitious than this—aiming for the protection of 80–90 per cent of the world's biodiversity, or even aiming to roll back the tide and restore biodiversity that has already been lost—the issues of global justice that I investigate will be yet more important. But either way, they are deserving of serious and urgent attention.

Conservation politics and global justice

That there *is* a global justice dimension to conservation politics should be obvious, even if that fact is sometimes obscured by the dominant

---

[12] Stark et al. (2021: 1) suggest that estimates of the required protection 'range from 28% to 80%, depending on the desired outcome'.

technical/scientific framings of the problem. Some recent proposals for responses to the biodiversity crisis would involve placing large areas of the planet beyond the reach of human 'development'. If so, we should expect to see serious constraints on some people's ability to escape from poverty or inequality. But actually, *any* robust and large-scale conservation policies appear certain to impose burdens—and probably very significant burdens—on the shoulders of some people rather than others. Rather than letting those burdens fall wherever they happen to fall, political theorists should offer guidance on where conservation burdens *should* fall, and what *just* conservation policies would look like. Decades of conservation practice have demonstrated that there is a significant danger that conservation policies will in practice compound existing inequalities, visiting still more injustice upon the poor and excluded. If policy makers and practitioners want to avoid those outcomes, they must ask: what would just conservation policies look like? By bringing the challenge of just conservation closer to the heart of debates about global justice, this book aims to make progress in tackling that question.

To date, there has been surprisingly little sustained effort to think through the global justice dimensions of the biodiversity crisis, and of countries' collective responses to it. There is, of course, a rich and extensive theoretical literature on the nature and value of conservation, which for the most part has been located within discussions of environmental ethics. But global justice scholars have not fully recognized the significance of the topic. This is in stark contrast to their treatment of a number of other 'applied' issues. Recent years have delivered, for instance, vibrant literatures on the global justice dimensions of migration, trade, and global health. But we cannot say the same thing for biodiversity conservation. There is an especially noteworthy contrast here with climate change, an issue many prominent political theorists have reflected on profitably and at length (see e.g. Vanderheiden 2008, Caney 2010, Gardiner 2011, McKinnon 2012, Moellendorf 2014, Shue 2014). Among other things, these theorists have thoroughly addressed problems of fair burden sharing posed by the need to stabilize the earth's climate, the tensions and trade-offs between mitigation and pro-poor development, and many applied matters of policy and institutional design. But biodiversity loss represents a similarly grave threat to our societies, calling the prospects of present as well as future human generations into question. It is a matter for serious regret, therefore, that political theorists interested in questions of global justice (as well as overlapping issues of natural resource justice and territory, for that matter) have only recently begun to turn their attention to important questions about the conservation of the living world (for rare counter-examples, see Armstrong 2016, 2017a, 2019a, and Tan 2021).

To the extent that there *is* an evolving debate on the justice and injustice of conservation politics, it has largely taken place within distinct fields such as environmental studies, conservation biology, and political ecology, and to some extent within other branches of political theory (such as ecological and animal ethics). Conservation scholars have themselves come to recognize that a more explicit discussion of the global justice dimensions of conservation policy would be highly beneficial (Martin, Akol, and Gross-Camp 2015, Martin 2017: 111).[13] A lively debate on where conservation ought to take place, who ought to bear the burdens, and what principles ought to guide the design of conservation policies, *is* therefore under way—but it has largely passed unremarked within the political theory of global justice. This, I believe, is highly unfortunate. It would be arrogant, of course, to assume that theorists of global justice will have all of the answers to the difficult questions we face in formulating fair and effective conservation policies. But we can surely make headway on some of these questions, and to the extent that this is true, the relative neglect of conservation justice in the field represents a missed opportunity.

Among other things, this book aims to persuade those currently engaged in debates about global justice (as well as overlapping debates about territory and natural resources) that biodiversity conservation does indeed raise important, even urgent, questions of global justice. It is vital (to human well-being and to the interests of many other creatures) that the world's leaders respond in a sustained and coordinated fashion to the looming biodiversity crisis. But, as I show in Chapter 1, the precise nature of their responses could have a significant impact on patterns of advantage and disadvantage the world over. Global justice scholars ought to be well placed to contribute to evolving debates about what just forms of biodiversity conservation would look like. To put it simply, biodiversity conservation ought to be a mainstream part of debates about global justice in exactly the same way as issues such as trade, migration, health, and climate change already are. If radical conservation measures are required—as I believe is wholly plausible—they should be designed, wherever possible, in such a way that they avoid exacerbating existing distributive or political injustices. They might even help us to move towards a more just world. Showing how is an important task, to which this book seeks to contribute. But I also hope to inspire others to place these issues closer to the heart of their work.

---

[13] Adrian Martin has made perhaps the most sustained attempt to consider the justice dimensions of conservation policy to date. However, he is clear that the focus of his excellent book is largely empirical: it aims to capture what real-world actors have said about just conservation, and 'does not seek to develop a normative theory of justice' in conservation (Martin 2017: 11).

Structure of the book

Chapter 1 has two goals. First, it shows why biodiversity conservation raises very important questions of global justice: because global inequality drives biodiversity loss, because biodiversity loss causes global injustice, and because responses to biodiversity loss will produce further global injustice unless formulated carefully and with clear normative guidance. For all of those reasons, the biodiversity crisis should play a much more central role within debates on global justice than it has to date. Second, the chapter sketches several prominent views on justice towards non-human animals. Rather than choosing between them, the aim is to show that, on *any* of these views, we have further reasons for caring about the biodiversity crisis. For the crisis threatens not only human interests, but those of the many other animals we share the planet with.

Chapter 2 clarifies the concepts of conservation and biodiversity and offers a pluralist account of our (justice-based) reasons for biodiversity conservation, which is grounded in the interests of both humans and non-human animals, as well as in the intrinsic value of biodiversity. It also clarifies the relationship between biodiversity conservation and some other environmentalist goals with which it is sometimes conflated.

Chapter 3 addresses our first major question of global justice, which is how the burdens of biodiversity conservation ought to be shared among us. It defends an account that foregrounds the contribution to the problem and ability to pay principles, although it also shows how patterns of benefitting can be relevant when we come to appraise people's failure to meet their conservation-related duties.

How conservation burdens should be shared is one important question, but another is how we should conceive of those burdens in the first place. Chapter 4 assesses how one kind of conservation burden—opportunity costs—should be understood and shows that our position on this issue has major implications for global justice. It argues that the notion of opportunity costs must be moralized and explains and defends an egalitarian baseline for calculating those costs. Taking this baseline seriously would suggest that many real-world conservation projects send far more modest resources in the direction of people affected by conservation projects than justice requires, and that they are often exploitative in nature. It examines the implications of that fact for conservation academics, practitioners, and policy makers.

Chapter 5 explores the role that offsetting can play in biodiversity conservation. It outlines the moral case for biodiversity offsetting, but it also shows that offsetting can serve to legitimate serious harms to both humans and non-human animals, and explores the worry that it can undermine commitment to more robust measures to preserve biodiversity. In light of these dangers, the remarkable proliferation of biodiversity offsetting schemes in conservation practice should be of serious concern.

Chapter 6 begins by discussing the Half Earth proposal—perhaps the most notable attempt to provide a large-scale, coordinated 'solution' to the biodiversity crisis. It offers an analysis of the moral costs that are likely to accompany any attempt to implement Half Earth, and suggests that the proposal, if enacted, would cause significant global injustice unless accompanied by robust side policies at the very least. This justifies a search for alternative policies that might deliver similar conservation outcomes, but with lower moral costs. At the end of the chapter, therefore, I identify an alternative package of policies that holds considerable promise. This package involves substantial legal protection for the land claims of indigenous and other marginalized peoples, and a set of structural reforms in the global economy— including debt forgiveness and the removal of harmful subsidies among other policies—which, taken together, could make significant progress in tackling the biodiversity crisis while helping, rather than hindering, wider projects of global justice.

Finally, the Conclusion draws together some the main arguments of the book in a reader-friendly format.

1
Biodiversity, justice, and animals

This chapter begins by showing why the biodiversity crisis demands serious reflection from anyone concerned about global justice—that is, anyone concerned about the duties and entitlements of human beings wherever in the world they happen to live. As such, biodiversity justice should be (though to date it has not been) a core concern within wider debates about the justice and injustice of the contemporary world order. Following that, it suggests that the interests of non-human animals *also* matter when we come to reflect on the biodiversity crisis. My aim here is not to be especially controversial—rather, I show how a series of influential positions on the moral status of animals can *all* underpin the conclusion that the biodiversity crisis poses a serious threat to the other animals we share the world with, and that these threats are morally weighty in their own right. This prepares the ground for the arguments of Chapter 2, and its pluralist account of our reasons for caring about the biodiversity crisis.

Why conservation and global justice?

Here I suggest that we have three powerful (but somewhat neglected) reasons for thinking about biodiversity conservation and global justice together. First, global inequality appears to be a major cause of the biodiversity crisis. As such, there are grounds for believing that tackling inequality will be crucial if responses to the biodiversity crisis are going to be effective. Second, the biodiversity crisis is a cause of global injustice. Caring about global justice therefore gives us further reason for seeking effective solutions to accelerating biodiversity destruction. Third, if they are not formulated extremely carefully, collective responses to the biodiversity crisis can produce significant global injustices in their own right. Regrettably, there is no shortage of real-world cases in which policies aiming at biodiversity conservation have exacerbated existing injustices between people, or even produced new injustices. Poorly designed conservation policies can be distributively unjust,

Global Justice and the Biodiversity Crisis. Chris Armstrong, Oxford University Press.
© Chris Armstrong (2024). DOI: 10.1093/9780191888090.003.0002

insofar as they generate unfair distributions of benefits and burdens. They can also be politically unjust, by paying excessive heed to the voices of the advantaged and excluding the voices of people with a good claim to participate in their design and implementation. The lesson to draw from this is that we cannot assume that biodiversity conservation and global justice are complementary projects which can be pursued together without any danger of conflict. To the contrary, policy makers need to work hard to *ensure* that global conservation policies are justly designed and implemented, as well as just in their effects.

Inequality as a driver of the biodiversity crisis

A number of empirical studies have suggested that distributive inequality is a key determinant—perhaps even *the* key determinant—of patterns of bio-diversity loss, either across all taxa (Mikkelson et al. 2007, Holland et al. 2009, Hamann et al. 2018), or across specific biological groups such as vascular plants (Pandit and Laband 2009). According to one overview, we have strong evidence that more unequal societies, for instance, experience greater environmental degradation, and that while the effects hit poor and ethnic minority communities disproportionately hard, ultimately the impacts are felt by all (Cushing et al. 2015). More recently, the Intergovernmental Science-Policy Platform on Biodiversity and Ecosystem Services has confirmed that more unequal societies experience greater rates of biodiversity loss (IPBES 2019). Various explanations have been suggested for this finding. One suggestion is that citizens in more unequal societies strive harder for conspicuous consumption, in contrast to more equal societies where less emphasis is placed on hierarchies of status (ibid: 196–7). It has also been suggested that gender inequality drives worse conservation outcomes, with more gender-unequal societies having higher per capita ecological footprints (Billé et al. 2013). Alternatively, it might be that highly unequal societies witness greater resistance to conservation policies, perhaps because social trust is lower, or because people do not feel that the costs of conservation are being allocated fairly (Martin 2017: 37, Hamann et al. 2018: 70). The suggestion here is that, in such societies, the advantaged are able to opt out by pursuing private consumption as opposed to making a fair contribution to protecting important public goods (Holland et al. 2009: 1312). Political inclusion, by contrast, has been said to make a positive difference to conservation outcomes. One study has found that as inequalities in political power diminish, so do pollution and other environmental problems

(Clement and Meunie 2010). If so, this may be because citizens can better agitate against environmental destruction in more inclusive political environments. In short, more equal and inclusive societies appear better able to mount the collective action required for environmental protection. High inequality 'tends to concentrate economic power within a select few who can circumvent rules, impeding efforts to halt biodiversity loss', as well as eroding institutional capacity (Mirza et al. 2020: 39). Reducing inequality, by contrast, appears to help make conservation rules bind. As a result, more equal societies are said to have lower levels of species loss, for example (Mikkelson 2013: 435–6).

Though these empirical claims are intrasocietal in scope, we have good reason to believe that inequality at the global level is also a significant driver of biodiversity loss. Adherents of the 'environmental Kuznets curve' (see e.g. Dinda 2004) might suggest that with greater wealth, biodiversity loss slows or even reverses. If that is true, then when some countries become wealthier than others, we might expect this to diminish overall biodiversity loss, rather than bolster it. But we should treat such claims with caution (Stern 2004). There is actually very little evidence pointing to the existence of a biodiversity-related Kuznets curve (Mills and Waite 2009). There are certainly glaring counter-examples: Australia and the US both make it onto the list of the seven countries with the greatest amount of biodiversity loss over recent years (Waldron et al. 2017). Other wealthy countries might not today be the site for extensive destruction for the simple reason that so much domestic biodiversity has already been lost. For example, Maron et al. (2020) suggest that countries like France and Italy possess less than 5 per cent of their 'original natural ecosystems' (a category we should of course treat with some caution), while the UK has been described as the most biodiversity-depleted country in the world.[1]

Perhaps most importantly, however, analyses of the connection between wealth and environmental destruction often focus narrowly on where destruction happens, and frequently draw the implication that it is locals who are responsible for it. But processes that degrade biodiversity in the global South are often linked to unsustainable lifestyles in the global North, and if they help to deflect attention from that fact, claims about a biodiversity Kuznets curve are not only suspect, but also genuinely damaging. Much of the soy grown on razed rainforests, for instance, is used to produce beef for export. Almost 40 per cent of the beef produced in the Amazon is shipped to

---

[1] https://www.nhm.ac.uk/discover/news/2020/september/uk-has-led-the-world-in-destroying-the-natural-environment.html (Accessed 3 February 2023).

the EU for consumption (Venter et al. 2016: 5). Mexican spider monkeys are endangered chiefly by a coffee trade oriented towards consumers in wealthy countries (Lenzen et al. 2012: 111). More generally, the proportion of mammals that are endangered in a country has been linked to its dependence on primary natural resource exports (Shandra et al. 2009), which typically involves servicing demand in more affluent parts of the world. Meanwhile, to a significant extent, the overfishing of the world ocean is driven by the activities of relatively few keystone corporations, which wreak an outsized impact on marine biodiversity (Österblom et al. 2015). Despite the heavy nutritional dependence of many communities in the global South on fish, these corporations oversee a large net transfer of landed fish from the global South to the North (Swartz et al. 2010: 1367). Overall, exogenous demand serviced by international trade has been connected to at least 30 per cent of global species threats, with Northern countries the main net importers of most of the commodities linked directly to biodiversity loss (Lenzen et al. 2012).[2] Biodiversity loss is also partly caused by other (highly unequal) ecological impacts such as climate change, which on a per-capita basis people in the North bear disproportionate responsibility for (Austin 2021).

It is important to acknowledge, then, that many of the countries with the highest rates of biodiversity destruction today are in the global South: alongside Australia and the US, countries like Indonesia, Papua New Guinea, Malaysia, and India are now among the leading locations for destruction of biodiversity (Waldron et al. 2017). This is partly explained by the fact that it is in the global South that most biodiversity can be found. But destruction in those countries is strongly linked to consumption in the global North. In recent decades, the world has witnessed a process of 'environmental load displacement' (Hornborg 2006) whereby the ecological demands of Northern lifestyles are increasingly placed on biomass located in the South. But to tailor conservation policies solely at preventing destruction at the local scale will often be point-missing and even counter-productive. Too often, a focus on proximate causes 'has led to the formulation of solutions that are simplistic with no lasting ecological benefits at best, and often downright unjust at worst, such as arming guards with shoot-to-kill powers in protected areas' in the global South (Pascual et al. 2021: 569). The problem with such policies is that they 'deflect attention from deeper, structural processes' at the global level (ibid). The fact that biodiversity loss is often bound up in socioeconomic processes at the global scale matters, that is, when we come to think

---

[2] Sometimes the impact of trade on threatened species is more direct, with many species caught to order and sold across borders. In such cases, species rarity can itself be a significant driver of rates of capture and sale, leading to a vicious spiral towards extinction (Hall et al. 2008).

about remedies. An advantage of thinking about conservation in tandem with global justice is that wider structural causes can be recognized, and remedies put in place at the appropriate scale. In the book's Conclusion I argue that the most just and effective responses to the biodiversity crisis would likely involve a shift away from coercive policies aimed at reducing the most proximate causes of biodiversity loss, in the direction of wider reforms aiming to reduce harmful subsidies, indebtedness, and the unsustainable consumption patterns of the advantaged, while simultaneously protecting, rather than undermining, the land claims of indigenous and other local communities.

The biodiversity crisis as a cause of global injustice

Scholars of global justice ought to be much more concerned than they have sometimes appeared about biodiversity loss, if for no other reason than that it promises to exacerbate existing global injustices. Since they tend to have weaker adaptive capacity, and because they are often under-represented in decision making about resource and land use, there are abundant examples of the least advantaged being hit hardest by the degradation of ecosystems (Díaz et al. 2006: 1302). For instance, environmental despoliation has been said to decrease the water and food security of the poor in particular (Hamann et al. 2018: 63). Within agriculture, the loss of biodiversity has left many poor people in the global South more vulnerable to crop failures (Ebel et al. 2021). Biodiversity loss reduces the safety net enjoyed by poor people who will often fall back on 'wild' food sources in times of hardship (Timmer and Juma 2005: 28). Climate change, meanwhile, is causing major geographical movements of fish populations and, in many places, a loss of marine biodiversity overall, an especial problem in the tropics where millions are nutritionally dependent on fish (Koenigstein et al. 2016). Finally, biodiversity loss and habitat destruction have been linked to the rising incidence of pandemics. As more and more forest or wetland is cleared, and more livestock animals are put in close proximity to it, the opportunities for viruses to jump the species barrier multiply (Lawler et al. 2021). It is often the world's poor who are most vulnerable to threats from new zoonotic diseases.

It is also vital to recognize that the world's poor are not a homogenous category. In many cases, women will be hardest hit by biodiversity loss because they are disproportionately dependent on the primary harvesting of local flora and fauna (Cripps 2022: 63). To the extent that biodiversity loss will hit people in the global South hardest, this means that it will often hit black people harder than white people. *Within* countries in the global South, meanwhile, people who belong to excluded and disadvantaged minorities may be especially badly exposed to threats to their basic interests. Within the countries of the global North, meanwhile, ethnic minority citizens may be disproportionately exposed to the effects of biodiversity loss, just as they have been disproportionately exposed to the harms attendant on climate change (Schell et al. 2020). The same can be said about indigenous people in the countries of the North, where these populations are far more likely to be dependent on direct harvesting of biodiversity than their fellow citizens, and who may be especially vulnerable because of ongoing problems of political exclusion and under-representation.

Finally, it is important to recognize that the lack of access to diverse ecosystems can constitute an injustice in its own right. As I note in Chapter 2, access to diverse and healthy ecosystems can be an important driver of individual well-being. In many places, however, ethnic/racial as well as economic cleavages translate into divergent access to green spaces (Dai 2011)—a trend that became especially visible at the onset of the Covid-19 pandemic. This in turn can impact negatively on the mental health of children and adults (McCormick 2017). Within inner cities, poorer children face inferior access to diverse green spaces, whereas the privileged can often take daily access for granted (as well as having greater capacity to engage with spectacular wildlife in their role as tourists).[3] Outside of cities, by contrast, poor people in the global South are often actually advantaged by readier access to diverse ecosystems, given their more extensive dependence on primary harvesting from local ecosystems. But this also means they can be especially vulnerable to processes of habitat destruction. This destruction will often have a transnational character, in a global economy where corporations seek out jurisdictions where constraints on environmental destruction are weakest.

Conservation policies as a cause of global injustice

As discussed, a growing number of scholars and activists have turned in recent years to radical and ambitious proposals aiming to salvage the surviving portions of relatively undisturbed biodiversity from the wreckage of human impact. According to one prominent call to arms, 'immediate efforts . . . of unprecedented ambition and coordination' will be required to turn back the tide of biodiversity loss (Leclere et al. 2020: 551). Most famously, the 'Half Earth' movement has called for the world to be divided

---

[3] A 'luxury effect', whereby access to biodiversity tracks affluence, has been repeatedly observed within cities in both the global South and the global North. For a review of the evidence see Leong et al. 2018.

into two halves, with one half given over to human exploitation, and the other returned to 'nature' (Wilson 2016). Alternatively, a Global Deal for Nature would involve the protection of half of all terrestrial ecosystems (Dinerstein et al. 2017). In addition, some scholars have called for the abandonment of the legal principle of permanent sovereignty over natural resources, and its replacement, at least when it comes to key ecosystems, with a system of trusteeship that would see significant constraints on what local states can do with them (Mancilla 2022).

Such proposals raise major questions of global justice. One of them is who will reap the benefits of conservation. Assuming that biodiversity is to be preserved in a more thoroughgoing way, who, if anyone, will be able to access it, or benefit from having it nearby? Many of the most pressing questions, however, concern the *burdens* of conservation, and these are my primary focus in this book. If governments were to undertake radical conservation policies, this would involve many people taking on costs. It may well be that radical conservation policies will end up being net-beneficial for human beings as a whole, especially if they rescue us from ecological disaster. Future people in particular are very likely to benefit from the success of such schemes. But in the short to medium term, such policies could seriously constrain what many people could do, where they could live, and how they could work. Even if those costs were outweighed in the long term by the benefits of living in a safe and bountiful biosphere, such policies seem quite likely to involve *some* people becoming worse off, at least within the current generation.

How should people decide which parts of the world are to be conserved, and which will remain open for extractive activities? If it transpires that conservation activities are not evenly distributed across the globe, will efforts to mitigate biodiversity loss place serious limits on some people's access to well-being, compared to others? When and how should we attempt to compensate or even avoid such costs? An account of justice in conservation ought to provide guidance on those and many other questions. Though the ambition of the radical conservation proposals mentioned above is to be welcomed, without serious attention to such questions there is potential for deep conflict between conservation practice and erstwhile goals of global justice. The challenge is made especially acute, of course, by the fact that much of the world's remaining biodiversity lies in the global South. This is in large part determined by the 'latitudinal biodiversity gradient', which sees a considerable decrease in biological complexity the further one moves from the equator (see e.g. Hillebrand 2004). Though the causes are somewhat contested, the fact that the world's 'biodiversity hotspots'[4] are disproportionately concentrated in the tropics is not. Although they make up only 6 per cent of the world's land surface, tropical forests, for instance, contain 50–90 per cent of all species (Youatt 2015: 32–6). Many of these species are endemic, meaning that they are found nowhere else on earth. Poverty, however, is *also* highly concentrated in the tropics, producing an extensive overlap between poverty and biodiversity (Barrett, Travis, and Dasgupta 2011, Maron et al. 2020). Insisting that tropical ecosystems such as rainforests are preserved intact could have serious ramifications for hundreds of millions of people desperately trying to escape from severe poverty. Thirty-five per cent of terrestrial areas with 'very low human intervention' are owned or occupied by indigenous groups (IPBES 2019: 14), and biodiversity is declining less rapidly in those areas than elsewhere (ibid 31). As a result, their members are likely to be disproportionately affected by coercive conservation policies.

Crucially, large-scale conservation policies arise against a context shaped significantly by colonialism and territorial dispossession. Colonialism (and the emergence of plantation economies) was often the cause of massive biodiversity destruction in the global South, but colonizers nevertheless repeatedly presented themselves, however improbably, as biodiversity's saviours, and locals who were dependent on subsistence harvesting as its greatest threat. Reflecting on the London Conventions on flora and fauna of the first half of the twentieth century, Rachelle Adam usefully reminds us that 'International biodiversity law did not start out as law between or among sovereign states addressing the biodiversity of all. Rather, it was originally colonial law imposed on Africa by colonial conquerors' (Adam 2014: 10). The depiction of poor locals as the real threat to biodiversity—conveniently deflecting attention from the consequences of Northern over-consumption—has roots at least a century old. In the current day, conservation politics continues to raise the spectres of territorial dispossession and neo-colonialism. Conservation projects in the South have in some cases led to indigenous peoples being forced from their traditional homelands, including at gunpoint (Peluso 1993). Since the 1990s, a self-proclaimed 'war for biodiversity' has sometimes been 'used to justify highly repressive and coercive policies' in the global South (Duffy 2014: 819), and has involved the widespread use of private military companies and the adoption, on occasion, of shoot-on-sight tactics targeting 'poachers' (ibid: 832) as part of a broader militarization

---

[4] A term invented by the tropical ecologist Norman Myers in 1988, and which has typically been taken to denote areas with both a high level of species diversity (especially of endemic species) and a high level of threat. See Youatt 2015: 21.

of conservation.[5] Chapter 6 discusses how the 'fortress conservation' model holds that the presence of 'local' people is an impediment to conservation that needs to be minimized (Brockington 2002). At the same time, in many places in the global South conservation policies have empowered and enriched unrepresentative leaders. Scholars of 'green colonialism' (e.g. Zaitchik 2018) suggest that an 'extractivist' model of economic development, in which the interests of indigenous communities and other disadvantaged groups were systematically disregarded, and where influence and benefits flowed to a privileged minority, has now been partly overwritten by a model of conservation that appears to work in much the same way.

There is a serious irony here. In the 1970s and 1980s, the 'environmental justice movement' arose to challenge the way pollution and environmental destruction disproportionately affected black, minority, or indigenous people both within countries like the US and in the global South (Schlosberg 2007, Part II). In decisions about where to site pipelines or dams, the interests of the socially and politically marginalized were often systematically discounted, leaving them disproportionately vulnerable to threats to their health and wider well-being. But *conservation itself* has also threatened to set back the interests of indigenous and other disadvantaged communities, and conservation decision making has disregarded their interests all too frequently. In the past, leaders who have invoked the idea of environmental crisis have gone on to steamroller democratic safeguards, hoarding influence and resources for themselves (Niemeyer 2014). The idea of a biodiversity crisis must not become a vehicle for further injustice, whether of the distributive or political variety. But ensuring that it does not is a major challenge.

How should we respond to the fact that conservation policies often threaten to visit further injustice on the poor and excluded? We should not allow valid worries about the injustices of some conservation policies to derail the project of biodiversity conservation altogether: after all, the biodiversity crisis will potentially threaten human and non-human life on a massive scale, and as with climate change, those who are already disadvantaged will be the most vulnerable. To the contrary, this litany of unjust policy shows why we must take the challenge of just conservation very seriously indeed. There is no reason to suppose that effective conservation policies will be distributively unjust and politically exclusive by definition. Policies that are both effective and just ought to be accessible to us. However, there is considerable work to do in formulating them—a task to which scholars

---

[5] For Duffy et al. (2019: 66), the militarization of conservation is characterized by the adoption of more forceful or armed tactics; the development of military-style and counter-insurgent strategies; and the use of technology originally developed for military purposes.

of global justice can contribute. One rather simplistic counter-move would be to assert that conservation policies must be just in order to be effective (see Martin, Akol, and Gross-Camp 2015: 167). If so, any apparent tension between effectiveness and justice dissipates. Certainly proposals that load conservation burdens onto the shoulders of the poor have often turned out to be self-defeating, since given the choice between escaping poverty and preserving biodiversity, many poor communities opt for the former (Sanderson and Redford 2003). Unjust conservation decision making has often sparked conflict, undermined trust, and hampered biodiversity preservation (Pickering et al. 2022: 164). Nevertheless, unjust conservation policies—which have excluded locals from participation, and even removed them from their traditional homelands—have on occasion been effective in practice (Brockington 2004), albeit sometimes at great social and financial cost (Dawson et al. 2021: 8–9). Participatory management approaches may be effective more often than exclusionary ones (Porter-Bolland et al. 2012, Oldekop et al. 2016), but exclusionary polices do sometimes deliver on their goals. Rather than being inevitable features of effective policies, fairness and inclusion need to be argued for, and need to be designed into conservation policy, from the outset.

This also means we should resist any claim that projects of conservation and global justice ought to be pursued separately or in isolation. For example, Kinzig et al. (2011: 604) suggest that the urgency of dealing with the biodiversity crisis means that social objectives such as poverty alleviation ought to be placed to one side. Though important in their own right, they declare that such objectives are best dealt with separately because hardwiring them into conservation policies will only lead to further delay. One problem with this view is that unjust conservation policies are often counter-productive, as we have seen. Another is that side policies will not always represent an adequate response to the injustices that conservation policies can engender. People are often deeply attached to the places where they live, the distinctive ecosystems they share their lives with, and the specific pursuits that support their livelihoods. Uprooting them from those places and attempting to remedy their losses by way of financial compensation, say, will be a distinctly second-best solution, if it is a solution at all, as will policies that offer retraining to compensate for the loss of livelihoods. There may occasionally turn out to be cases where such policies are ultimately necessary, in the sense that effective conservation is simply not possible without them, and where the failure of conservation will unavoidably thwart the most basic interests of even more people. Typically, however, such policies will *not* be necessary in that sense. To uproot people from their homes and disrupt their most cherished projects, when there are accessible and effective policies that would not have that out-come, would represent an unjustified assault on their autonomy. It would also show disrespect for the life-plans they are committed to (in Chapter 6 I exam-ine related worries about Half Earth proposals). If mainstreaming a concern for justice from the policy formulation stage means that such outcomes can be avoided, then they ought to be. Thinking through what just conservation policies might look like ought to be a priority.

Beyond anthropocentrism

So far in this chapter I have concentrated on issues of global justice. I take that term to refer to the proper distribution of benefits and burdens between individual human beings. If left unchecked, biodiversity loss is likely to have many negative—perhaps even catastrophic—consequences for members of our own species. Concern for those consequences is sufficient, I believe, to justify a shift from the mere description of biodiversity *loss* to the moralized language of a biodiversity *crisis*. To the extent that we can avoid dangerous outcomes for individual human beings, then citizens—and the collectives they make up—can be under a duty of justice to do so. Since theories of global justice assess how the benefits and burdens of living together on a lim-ited planet ought to be shared between us, theorists of global justice ought to be interested—much more interested than they appear to have been to date—both in the impacts of biodiversity loss, and the impacts of collec-tive responses to it. They also ought to be interested in the intergenerational dimensions of the problem, on the assumption that most of the dangerous effects of unmitigated biodiversity loss are likely to affect people living in the future.

So far in this chapter, however, I have not referred to non-human interests. For anthropocentrists about justice, the discussion about the biodiversity cri-sis could stop here. To be more precise, anthropocentrists might recognize that the biodiversity crisis will *also* be very bad for non-human animals. They might even argue that we are morally obliged to mitigate or prevent some of its negative consequences for the other animals with whom we share the world. But if they did, they would insist that these obligations were not duties of justice. Justice and injustice are terms, on this view, that we use to refer to relations—including relationships of entitlement and obligation—between human beings alone (see e.g. Rawls 1971).

The danger the biodiversity crisis poses to (current and future) humans and their interests clearly does provide powerful reasons for seeking to arrest or even reverse it, and this suffices to make the inquiry into the global justice dimensions of biodiversity conservation vitally important. Nevertheless, it is unclear that anthropocentrists can give a fully satisfying account of either the nature of the biodiversity crisis, or of our duties in virtue of it. As I argue in Chapter 2 (and as I have argued elsewhere: see Armstrong 2022, chapter 7), it is not only humans who possess important interests, including, more specifically, interests that are weighty enough to place burdens of justice on the shoulders of others. Individual non-human animals do, too. If so, we are pushed in the direction of a non-anthropocentric account of justice. Anthropocentrists, to be clear, can accept that we possess duties *with regards to* non-human animals, if treating them badly turns out to have important implications for human well-being. Immanuel Kant's argument that people should not mistreat animals—because people who do so are likely to go on to treat humans badly, too—would be an example of such an argument.[6] Non-anthropocentrists, by contrast, are happy to accept that humans can owe duties of justice *to* non-human animals, in virtue of the latter's interests. On this view we should refrain from mistreating animals not only because doing so makes one more likely to mistreat humans, but also because mistreating animals is unjust in its own right.

What difference does a recognition of animals as subjects of justice make to a discussion of the global justice dimensions of the biodiversity crisis? To reaffirm a point I made earlier, when I speak of global justice, I have in mind the ways human beings share benefits and burdens together. This is not to deny that any view about global justice ought ideally to be nested within a broader account of ecological justice, which will potentially specify how benefits and burdens should be distributed between all living beings. But my aim here is not to set out an account of ecological justice. Rather, it is to address the (inter-human) global justice dimensions of that crisis, and of our collective (human) responses to it. Why, then, make this foray into discussions about animals and justice at all?

The answer is that the interests of non-human animals matter to my inquiry in at least two distinct ways. First, while concern for the interests of humans is undoubtedly sufficient to motivate serious attention to the biodiversity crisis, and serious reflection on its global justice dimensions, our priority should not always be to provide a merely sufficient account. In fact, there is a danger that focusing on merely sufficient reasons for addressing the biodiversity crisis will give us misleading guidance on its importance, when

---

[6] Kant also believed that people who mistreat animals might be violating a duty to themselves by undermining their own capacity for empathy (see Kant 2017: 6: 4333). But this, too, is compatible with anthropocentrism about justice.

compared to other political priorities. Recognizing the ways in which animals' interests are bound up in the biodiversity crisis can be important when we decide how much time, effort, and resources to devote to addressing it, compared to other goals. Neglecting important reasons for caring about a political cause is likely, by contrast, to give a distorted sense of its relative priority. A focus on human interests tells us that the biodiversity crisis could have very bad outcomes. Acknowledging the way in which the interests of individual animals are also bound up in the biodiversity crisis tells us that those outcomes would be still worse than that. As such, the interests of animals are likely to make a difference when we come to address the biodiversity crisis, just as they can when we come to address other moral issues, such as the conduct of war (Milburn and Van Goozen 2020). Second, taking the interests of non-human animals into account can also matter when we come to evaluate particular biodiversity-related policies. In Chapter 5, I argue that the practice of biodiversity offsetting threatens to license harms to many animals. If so, this must matter when we come to evaluate that practice. Ignoring that impact might lead us to consider policies to be permissible when in fact they are not.

Three views on the status of animals

In this book I do not rely on any claims about the rights of, or our duties of justice towards, plants. Neither do I investigate purported duties of justice towards species or ecosystems considered as collectives. Instead I restrict my focus to zoocentric individualist views, which suggest that (only) individual animals can be the bearers of entitlements and, potentially, duties (see e.g. Donaldson and Kymlicka 2011, Nussbaum 2022). Some views on animals and justice would add further qualifiers, suggesting that it is *sentient* animals, say, who can possess entitlements of justice (e.g. Nussbaum 2022, chapter 6). I am not persuaded of that: it seems more likely to me that while sentience can explain why a being is the right kind of entity to have entitlements, so too can other features such as subjecthood, or autonomy. Very many animals will turn out to possess at least one such feature. The key point, though, is that while biodiversity loss can matter to human interests, it can matter to the interests of (very many, if not all) non-human animals too. As I argue in Chapter 2, this leads to an expanded account of our reasons for addressing the biodiversity crisis, and of the ways in which we should address it.

However, in cases where the interests of humans and other animals come into conflict, zoocentric views diverge significantly on the question of whether we should weigh human interests more heavily than those of other animals. Consider these three different views, or types of view, which reflect some of that diversity.

First, those who defend 'hierarchical' views about justice believe that the interests of animals count for *something* from the point of view of justice, but not for as much as those of humans, because while animals have some moral status, it is lower than that of humans. In principle, the hierarchy in question might be very steep, so that the interests of humans count for very much more than those of other animals, simply because they are human interests. On such a view the interests of animals could even end up counting for almost nothing. More plausible would be a 'moderate' hierarchical view, according to which human interests count for somewhat more, but are not weighted extremely heavily compared to those of non-humans. According to Shelly Kagan, for example, 'Animals count for less than people do, but they count for far, far more than we ordinarily acknowledge' (Kagan 2018: 17). Although Kagan does not explicitly adopt the language of justice, his view would appear to have strong implications for the permissibility of various political policies, and for the kinds of rules that can—and should—be coercively imposed.

Second, those who defend 'egalitarian' views believe that the weight we give to interests should not depend on species membership. One variant of the egalitarian view (which is often associated with utilitarianism) holds that all animals have equal moral status, and therefore suggests that *equal interests count equally* regardless of species membership (Singer 1974). On that basis, causing hunger or pain for a mouse is as bad as causing hunger or pain for a human. This view does not hold, however, that the *overall* well-being of individual members of different species is to be given equal moral weight. Adherents of the view might say, for instance, that a human is likely to have *more* interests at stake than a mouse when policy makers come to make decisions about resource use. It might also be said, on this view, that the death of a human is much worse than the death of a mouse, because a human has far more interests at stake in continued life than a mouse does. The view simply insists that where the *same* (or a very similar) interest is held by members of different species, those interests ought to be weighed equally. The alternative is to commit the sin of speciesism: to assume that a given interest should be treated as weightier *simply because* of the species to which its holder happens to belong. Speciesism in that sense appears to be objectionable in much the same way that racism and sexism are (Singer 1974). A more demanding egalitarian view would hold that the total well-being of each animal ought to be seen as equally weighty when decision makers come to distribute resources (see e.g. Holtug 2007, Horta 2016). This view would rule out directing more resources towards a human simply because he or she had more interests at stake than a mouse. Instead, it would require decision makers to direct resources to whoever is worst off as a whole. If we consider the well-being of non-human animals to be generally lower than that of humans, this brand of egalitarianism would speak in favour of what Nils Holtug (2007: 10) has called 'a massive shift of resources from most humans to most non-human animals'.

Again, these egalitarian views do not necessarily adopt the language of justice, even if they do appear to have powerful implications for political policy. A third view, by contrast, is explicitly expressed as a statement of what we owe by way of justice to other animals. Martha Nussbaum (2022) argues that her capability theory should be extended to all sentient non-human animals. Nussbaum aims to set out a framework for assessing whether institutions and laws are just or unjust, and a key part of answering that question involves assessing whether they protect the capabilities of animals, or whether they thwart their distinctive forms of flourishing. The capabilities of different animals will vary partly (if not wholly) in line with their species membership. For each individual animal, there will be a list of core capabilities that ought to be in place, and that will involve rights to, *inter alia*, a healthy environment, freedom of expression, freedom of movement, and the like. It is not immediately obvious how we should categorize this third view alongside its hierarchical or egalitarian rivals. Like the egalitarian, Nussbaum believes that it is wrong to give less emphasis to an animal's interests simply because it belongs to another species. But, in line with her aim of securing overlapping agreement on legal and political essentials, Nussbaum focuses on thresholds for minimal justice, including the protection of *basic* forms of flourishing. On such a view, it is not obvious that humans commit an injustice if they protect the basic rights of non-human animals but use the resources available to them to advance their own interests rather more substantially. In terms of practical demandingness, it could be that the theory therefore ends up falling somewhere between the two first views. But I do not seek to settle this question here.

Neither do I seek to choose between these different zoocentric views. I simply want to note that *all* of them have significantly revisionist implications for political practice, including global conservation efforts. Advocates of the moderate hierarchical view, for instance, can hold that 'Our treatment of animals is a moral horror of unspeakable proportions, staggering the imagination' (Kagan 2018: 6)—and in need of urgent change. More broadly, defenders of hierarchy—whether of a deontological or consequentialist hue—can agree that non-human suffering is massive, relatively neglected, and often tractable in practice, and that reducing it should be one of our very highest political priorities (Sebo 2021). They can also accept all of the reasons for conservation that I outline in Chapter 2. Advocates of either egalitarian view will place even higher priority on ensuring that non-human animals can lead good lives, and this will plausibly have even more demanding implications for conservation practice. Nussbaum, for her part, accepts that the protection of both biodiversity in general, and species in particular, can be important to the interests of individual animals (Nussbaum 2022: 112). All three views, in short, have important implications for the kind of conservation politics we ought to pursue; they will plausibly help determine both how ambitious our conservation policies ought to be, and what kinds of policies or interventions we can legitimately take up in pursuing conservation goals. In Chapter 2 I simply assume a principle that all of these zoocentric individualist views hold in common, which is that the interests of non-human animals count for *something* at the bar of justice, and that this something is far from negligible. This helps to explain, I believe, why policy makers should make the biodiversity crisis one of their highest political priorities.

2

Theorizing biodiversity conservation

This chapter begins by clarifying the concepts of conservation and biodiversity. It then sets out a pluralist account of our reasons for caring about the biodiversity crisis. Finally, it situates the concern for biodiversity within the broader project of biological conservation. Although biodiversity is not the *only* thing that matters when it comes to protecting the living world, it is an important *part* of what matters. This gives us abundant reason for concern about the extent of the biodiversity crisis.

Defining conservation

In a very general sense, acts of conservation seek to defend or protect entities that have value, or which ought to be valued, so that they can continue to exist into the future (Cohen 2013, chapter 8). Familiar examples include efforts to protect ancient buildings or statues from the ravages of acid rain or erosion by the elements, so that future generations can experience them just as we have. Such efforts are defensive in character, insulating these artefacts from threats that will damage or destroy them if left unchecked. Threats can conceivably be human or non-human in origin (or a combination of both). An example of conservation in the face of a 'natural' threat would be moving statues from the path of a volcanic eruption. But many or most threats are likely to be anthropogenic in character, or largely anthropogenic. In 1959, for example, the government of Egypt proposed to flood the valley containing the Abu Simbel temples, dating from the time of Ramses II and widely regarded as a treasure of human civilization. The project to conserve the temples—which were eventually relocated further from the dam—was an early example of an internationally coordinated conservation effort. International concern about the initial threat to the temples provided impetus to the emergence of the World Heritage Convention (though its gestation was a long one: the Convention would not be adopted until 1972).

Global Justice and the Biodiversity Crisis. Chris Armstrong, Oxford University Press.
© Chris Armstrong (2024). DOI: 10.1093/9780191888090.003.0003

Though these efforts to protect concrete artefacts are familiar, in principle we might say that the activity of conservation is also in play when people try to preserve less tangible entities such as languages, or even the distinctive character or ethos of particular institutions. But while the preservation[1] of cultural artefacts or traditions can undoubtedly raise interesting questions of justice in its own right (Fabre 2021), my focus in this book falls not on human-made cultural artefacts but rather on elements of the living world. Here, too, examples of purposeful conservation activity come readily to mind. The National Trust, one of the United Kingdom's (UK) largest membership organizations, aims (among other things) to conserve healthy and beautiful 'natural' spaces.[2] The Worldwide Fund for Nature (known in the US and Canada as the World Wildlife Fund) has more than five million supporters across the globe, and describes itself as the world's leading conservation organization. It aims to protect vulnerable habitats and to aid in efforts to save species from the threat of extinction (its website gives pride of place to primates, big cats, and marine mammals, though its efforts actually extend much more widely).[3]

The activity of conservation is, however, a rather complex idea, which involves not one but several types of action or inaction (see Armstrong 2017a, chapter 10). First, conservation can simply mean the conscious *avoidance* of certain options. Examples include refraining from building a road on the grounds that doing so would destroy a biodiverse habitat, halting the use of a specific pesticide because of its toll on bird life, and cutting back fishing activity to allow fish populations to recover. In each of those cases a conscious decision is made to desist from a pursuit that is appealing to some but that is understood to be deleterious to non-human life. Second, conservation can mean the active *protection* of valued entities from various threats. Examples include removing discarded fishing nets from the ocean, patrolling the plains of southern Africa to prevent the killing of rhinos or elephants, and building firebreaks in dry forests (I take no stance here on the merits of these various activities). In each instance there is a conscious effort to insulate non-human life from a significant danger, which could be either anthropogenic or non-anthropogenic in origin. Third, conservation can involve efforts aimed at

---

[1] A note on terminology. Some environmental ethicists (e.g. Passmore 1974) have suggested that conservation is an anthropocentric concept (referring, e.g. to maintaining stocks of things likely to be useful to humans in future), whereas preservation is non-anthropocentric (referring, e.g. to protecting things from human impact, in light of their intrinsic value). That usage has been traced back to the competing visions of John Muir and Gifford Pinchot in the US (see Murdock 2021). But I do not lean on any such distinction here: I use both conservation and preservation interchangeably and inclusively to mean any measures taken to protect, restore, or avoid threats to biodiversity.

[2] It also aims to preserve historic buildings. https://www.nationaltrust.org.uk/features/strategy (accessed 7 October 2021).

[3] https://www.worldwildlife.org/about/ (accessed 23 September 2021).

restoration, including projects aiming to recover previously existing biodiversity, or habitats hospitable to it. One example would be the captive breeding schemes that have successfully reintroduced the Arabian Oryx, the California Condor, and Przewalski's Horse to the wild. Another would be efforts to repair coral reefs damaged by ocean warming, or to establish new colonies of coral in places less vulnerable to its effects.

Does calling restoration a form of conservation stretch the latter concept to breaking point? If conservation means securing things that have value, then it might be thought that it cannot plausibly include efforts to bring *new* creatures or *new* habitats into existence. This worry can be assuaged, though, by pointing to a familiar distinction between types and tokens. Whatever their general merits, efforts to breed the Arabian Oryx in captivity do clearly aim at the birth of additional oryxes, who would not exist but for those efforts. But if the goal of the activity is for the *species* to become self-supporting in the wild, there are no great conceptual gymnastics involved in describing this as a conservation project. Likewise, the transportation of corals to new, cooler parts of the sea floor may not preserve any particular coral *reef*, but it could preserve coral reefs as a distinctive *type* of ecosystem, as well as particular *species* of coral. This corresponds with common usage within conservation biology, where practitioners define conservation as an activity involving both the maintenance and restoration of ecosystems or biodiversity more broadly (see e.g. Vucetich et al. 2018: 23). We might want to expand the language of conservation even further to include efforts to renew or revitalize ecosystems, and not merely to restore what once existed. If so, the objection may again be that including ideas such renewal or revitalization stretches the concept of 'conservation' too far. Perhaps it does, in which case some other heading would be needed under which to collect these efforts. But either way, the questions I pursue in this book about fair burden sharing, fair participation, and so on would still arise, and would still be worthy of investigation.

Conserving biodiversity

The term biodiversity was coined in 1986 by plant physiologist Walter Rosen as a contraction of the term biological diversity.[4] Initially it was very common to identify biodiversity exclusively with the prevention of species loss (Youatt 2015: 28). However, in recent decades it has become much more common to follow Elliott Norse and colleagues (Norse et al. 1986) in defining

---

[4] The fuller term 'biological diversity' was in use as early as the 1960s (Farnham 2017: 11).

biodiversity as a measure of variation at the three distinct levels of genes, species, and ecosystems (see, e.g. Barrett, Travis, and Dasgupta 2011, Burch-Brown and Archer 2017), all of which appear to be important to the health of our living environment (Dasgupta 2021: 53). It is this more complex usage that has pride of place within international law. The Convention on Biological Diversity defines it as 'the variability among living organisms from all sources including, *inter alia*, terrestrial, marine and other aquatic ecosystems and the ecological systems of which they are part; this includes diversity within species, between species and of ecosystems'.[5] Biodiversity conservation, then, can be understood as an activity that aims to reduce the loss of this variability and/or to promote its recovery.

Within the scientific literature on biodiversity, there is widespread disagreement about the best way of unpacking and measuring biodiversity (Faith 2017). Furthermore, it is difficult to arrive at any single overall scale of biodiversity, at least without privileging one of its elements (genetic, species, or ecosystem-based variation) over the others. But what are we to conclude from this? The fact that so many different ways of practically measuring biodiversity possess at least initial plausibility has led some to argue that the term is so capacious that we would be better off without it (Santana 2014). Certainly, there are formidable obstacles to developing measures of biodiversity that are on the one hand practically useful, and that on the other hand do justice to the complexity of the core concept. Above all, biodiversity measures need to be capable of communicating important changes to society and to policymakers (Wilson et al. 2017). For this reason, simplifying surrogates for biodiversity are often used in practice (see chapter 5). But none of this establishes that we ought to dispense with that concept (Faith 2017). Individual measurements of biodiversity can be compared according to their practical utility within conservation planning (Sarkar 2006). The general concept of biodiversity, meanwhile, can serve as a useful overarching category capable of providing a focus for many conservation efforts even if its components cannot readily be reduced to one 'core' scale or measurement (Burch-Brown and Archer 2017). The examples of well-being and poverty show that concepts can be important—even indispensable—while simultaneously being irreducibly multidimensional.[6] Indeed, the multidimensional nature of biodiversity sits comfortably with the idea that we have a plurality of reasons for valuing it; a pluralist account of the grounds of conservation will gain appeal

---

[5] United Nations Convention on Biological Diversity (1992), Article 2.
[6] See, for example, Hassoun et al. 2020 on the multidimensional nature of poverty.

in part because it can accommodate this fact. In the next section, I set out such an account.

Why care about biodiversity loss?

Some environmental scholars have attempted to debunk claims about the value of biodiversity, for instance by questioning whether biodiversity is *always* instrumentally valuable to humans, or whether it is smoothly or inexorably related to the vitality or resilience of wider ecosystems (see e.g. Maier 2013, Newman, Varner, and Linquist 2017). Such accounts are a useful corrective and highlight some of the empirical complexities involved in claims about the importance of conserving biodiversity. But these critical accounts fail to show that biodiversity loss cannot have highly negative consequences. More importantly, they do not undermine the broad scientific consensus that the sheer level of biodiversity loss today is already threatening the well-being of many humans, as well as the prospects of the other creatures that constitute it (IPBES 2019). In this section, I set out four reasons for caring about biodiversity loss. Each gives us at least *pro tanto* moral grounds for engaging in biodiversity conservation. Ultimately, accepting any of these suffices to make tackling the biodiversity crisis an important project. I believe that each reason has weight (even if I am somewhat cautious about the implications of the fourth reason), and that any successful conservation politics must pay heed to all of them.

Biodiversity sustains vital ecosystem processes on which humans depend for their most basic rights

The UN Special Rapporteur on Human Rights declared in 2017 that 'the full enjoyment of human rights . . . depends on biodiversity', whereas its degradation and loss undermines those rights (Human Rights Council 2017: 3). It is uncontroversial that humanity's survival depends upon the existence of a healthy living world. Healthy ecosystems sustain large-scale biochemical processes such as the oxygen cycle, the nitrogen cycle, and carbon sequestration, without which it is difficult to imagine human life as we know it persisting. In turn, most ecosystems depend on a complex and diverse web of net primary producers such as plants, algae, or phytoplankton, all of which use photosynthesis to harness energy from the sun. Crucially, there is wide scientific consensus that healthy ecosystems will typically be diverse.

For instance, various reports indicate that ecosystems exhibiting greater variation are healthier, as well as more resilient, in the face of change, and more capable of sustaining key biochemical cycles as a result (see e.g. Naeem et al. 2012, Cardinale et al. 2012: 60). By contrast, there is unequivocal scientific evidence that the loss of variation reduces ecosystems' ability to produce biomass and cycle nutrients, and that these diminutions in ecosystem abilities are nonlinear, accelerating as biodiversity loss increases (Cardinale et al. 2012: 61, Dasgupta 2021: 74). More diverse environments are on average more productive and robust than those that are more uniform. According to one survey of the literature, the evidence suggests that conservationists should focus here not simply on the number of species, but also on the genetic diversity of their populations (Díaz et al. 2006: 1301), although species diversity is also positively linked to the productivity of ecosystems (Dasgupta 2021: 69).

It may well be that a good deal of biodiversity could be destroyed without causing the widespread collapse of ecosystems (this, of course, is a hypothesis the world's economies are currently putting to the test). One possibility is that the connection between, say, species diversity and ecosystem functioning is non-linear and saturated, so that the world could lose a good number of species without that making much difference to the overall health or productivity of ecosystems (see Newman, Varner, and Linquist 2017, chapter 2). But even on such a view we will still, eventually, run up against limits to our ability to remove elements from ecosystems without seriously damaging their functioning. The idea of a biodiversity *crisis* points to a situation where destroying biodiversity leaves humanity open to the threat of tipping points and threshold effects that could undermine the basic environmental and food security of hundreds of millions of people—and perhaps all of us. If these effects came to pass, such changes would be likely to threaten the ability of many people to meet their most basic rights—leading, if unchecked, to waves of starvation, dehydration, and displacement. If ocean acidification were to cause a collapse in zooplankton populations, for instance, there would be dramatic implications for the food security of hundreds of millions of people. If it were to cause radical changes to phytoplankton populations, the results could disrupt major planet-wide biochemical processes and threaten ecosystems much more broadly (Sepúlveda and Cantarero 2022). There are also well-known concerns that the loss in insect diversity may threaten the food security of millions of people, as pollination of food crops is reduced (Van der Sluijs and Vaage 2016). An estimated four billion people, meanwhile, 'rely primarily on natural medicines for their health care' (IPBES 2019: 10), a dependence that could turn into dire need if biodiversity continues to be rapidly eroded. Alternatively, consider the fact that genetic biodiversity often has informational value, which can be used by those seeking to develop medicines to combat illnesses that afflict humans. Other things being equal, the more genetic information that is lost, the lower our chances of tackling such illnesses effectively in future (Deplazes-Zemp 2019).

There are various ways of capturing our dependence on key biochemical processes and, by contrast, our vulnerability to their systematic degradation. One framing, influential from the 1970s onwards, suggests that there are systemic 'limits to growth' on a planet with finite natural resources (Meadows et al. 1972). Another framing points towards the existence of 'critical natural capital', which will be required in perpetuity if key ecosystem functions are to continue in a robust fashion (Ekins 2003). In recent years, the idea of 'planetary boundaries' has come more to the fore. The concept of planetary boundaries aims to define a 'safe operating space' for humanity, although scholars highlight that we have already overstepped some of these boundaries—notably so in the case of biodiversity (Rockström et al. 2009). Alternatively we might focus on the earth's 'regenerative capacity', construed in terms of its ability to keep on producing the goods and services that humanity consumes. On one estimate, humanity's annual consumption grew from 73 per cent of the planet's regenerative capacity in 1960, to 170 per cent in 2016 (Lin et al. 2018), meaning that the world's economies are now steadily eroding their own ecological foundations. Despite their differences,[7] each of these framings is capable of capturing the key point that—even if there is some 'redundancy' built in—a substantial degree of biodiversity is a prerequisite for even the most basic forms of human flourishing, and that if unabated, the loss of biodiversity will ultimately threaten all of our interests.

A vital task for political theorists, of course, is to plot the path towards duties and entitlements of justice. In the case of our first reason for conservation, the argument should be straightforward. To avoidably deny someone the means of meeting their most basic rights will typically be an injustice (unless doing so is necessary, say, in order to avoid basic rights shortfalls for many more people). In that context, it is hard to see what human interest could possibly justify destroying the life-support systems of (at the very least) hundreds of millions of people. All of us have basic human rights to, among other things, life, health, and subsistence (Shue 1980). Just as climate change jeopardizes all of these (Caney 2009), so too does large-scale biodiversity loss. As with climate change, it is likely that the biggest impacts of unconstrained

---

[7] For a survey of various accounts of ecological limits, and their normative underpinnings, see Green 2021.

biodiversity loss will be felt by future people (see Cardinale et al. 2012). As a result (where basic human rights are concerned), it is primarily in the interests of future people that we ought to act to tackle the biodiversity crisis. Whoever they will be, future people will require a liveable ecosystem capable of supporting basic biochemical processes. The biodiversity crisis threatens that foundation, and this alone is sufficient to justify a commitment to robust conservation policies. But the crisis also threatens the most basic interests of many currently living people, especially in the global South. This, again, should be ample reason to drive the issue to the top of the agenda within contemporary politics, as well as within global justice scholarship.

The ability to interact with a diverse living world is important for flourishing lives

The first point focused on the ways in which diverse ecosystems sustain human life, and the damage the destruction of biodiversity can do to the most basic human interests. But the ability to access healthy and diverse ecosystems is important to our well-being even when our survival is not at stake: it is an important constituent, we might say, of a *flourishing*, rather than a merely adequate, life. In recent years there has been a wealth of empirical research into the benefits of access to healthy green and blue spaces for our mental and physical well-being, and the harm that environmental destruction can do to it (see e.g. WHO 2015, Britton et al. 2020). The salience of healthy ecosystems for our well-being can be accounted for by a number of different views on the 'currency' of justice. A defender of an objective list account of well-being, for instance, can readily accept that living within a secure and diverse environment is an important component of well-being that everyone should be able to take for granted.[8] Defenders of capability theories of justice, for their part, have argued that access to a diverse and healthy living world is an important capability that should be in place for everyone (Nussbaum 2022).

In addition, many defenders of capability or objective list accounts of well-being have argued that the ability to experience *beauty* is an important element of a flourishing life. While humans have certainly created wonderful artefacts of their own, there is great beauty to be witnessed in the living world, from the aesthetic features of particular organisms (Moellendorf 2014: 49–53), to the spectacularly intricate and interdependent assemblages

---

[8] The same would follow for more complex accounts that maintain space for a subjective component to well-being. See Wall and Sobel (2021) for a defence of what they call a 'robust hybrid' account of well-being (to which I am sympathetic).

to be found in ecosystems such as forests, rivers, coral reefs, and tidal zones. Holmes Rolston III (2012: 49) has argued that 'life would be impoverished with reduced experience of natural beauty, rural and wild'. This view suggests that those who are unable to enjoy the aesthetic wonders of the living world are deprived of something very important, and this offers support to claims that highly unequal patterns of access to biodiversity (on which see Wolch et al. 2014) can constitute an injustice.

In the case of aesthetic appreciation, our interests can be set back directly by the destruction of the living world. But it also appears that our interests can be set back indirectly by biodiversity loss. For instance, it is plausible that a flourishing life includes the opportunity to (justly) engage in parenting children. If the biodiversity crisis puts people in a situation where they cannot justly bring children into the world—because the basic rights of those children would be highly insecure—then people's ability to (justly) flourish as parents will also be curtailed (Gheaus 2019).

We potentially have a number of reasons, then, for believing that a flourishing life depends upon the existence of a rich and diverse living world. But while what I have said so far has been rather general, it is important to recognize that people often have important and meaning-conferring relationships with *particular parts* of the living world, in all of their diversity. In such cases, specific ecosystems or assemblages of species should not be viewed merely as interchangeable supports for well-being. To the contrary—elements of the living world can matter deeply to people in their specificity. In the language of the Intergovernmental Science-Policy Platform on Biodiversity and Ecosystem Services (IPBES), these elements have 'relational' value, which expresses 'the role of contextual bonds to places or practices' (Anderson et al. 2022).[9] Respecting these bonds means recognizing that uprooting people from places or ecosystems they are attached to would likely cause them to bear significant costs. For example, consider the significance to many Aboriginal people of the interior of the Australian continent, or the importance of particular reindeer-herding routes to the Sami people of northern Scandinavia. Indeed some people will have worldviews or cosmologies in which their lives simply make no sense when detached from wider ecosystems, and will reject any easy distinction between harm to biodiversity and harm to themselves. The point cuts two ways. Attachment to—and embeddedness within—particular places and particular ecosystems is an important fact that just conservation

---

[9] There is some controversy about whether relational value is a subcategory of instrumental value, or distinct from it (see e.g. Deplazes-Zemp 2023). I do not take a position on that here, and none of my substantive arguments seem to hinge on where we place relational value within these typologies.

policies will have to take account of. But it also helps explain why conservation activities can be important. The degradation or destruction of some species, populations, or ecosystems can represent a significant harm to the members of specific communities, thwarting their ability to maintain projects that matter deeply to them.

Our first reason for biodiversity conservation emphasized the important role it can play in safeguarding our very survival. When such safeguards are imperilled on a wide scale, this is enough to justify us in adopting the language of crisis. But our second reason for conservation can lend further support to that choice of terminology. When biodiversity loss threatens *additionally* to undercut many peoples' ability to lead flourishing lives and to pursue meaningful connections, this matters too, and ought to be recognized as an important part of any crisis, which makes it still more serious. As we have seen, there is ample reason to fear that human flourishing will be seriously imperilled if the biodiversity crisis is unabated. According to the famous 'planetary boundaries' approach, for example, there are nine boundaries which specify a safe operating space for humanity. Outside of those boundaries, the basis for civilization and society are likely to be imperilled— even if human life continues. According to scientists, we have already begun to cross two of these nine boundaries: those relating to biospheric integrity (especially when it comes to genetic diversity), and to biochemical flows (Steffen et al. 2015).

The importance of biodiversity to flourishing lives is relevant to a theory of global justice, and capable of underpinning enforceable rights and duties. Minimalist accounts of global justice suggest that our chief concerns at the bar of justice are to avoid causing deficits to others' basic rights, to avoid exploiting people's absolute disadvantage, and (at least in some circumstances) to offer positive assistance to people trying to escape from it.[10] Such theories should obviously take our first reason for conservation very seriously, and as a result ought to embrace *pro tanto* duties to avoid causing biodiversity loss whenever this is likely to jeopardize people's basic rights. But on minimalist theories, it is less obvious that there are automatic grounds for concern when an impaired environment prevents some people but not others from leading *flourishing* lives (in cases where their basic rights are not threatened).

An egalitarian account of global justice, by contrast, ought to show concern with people's *comparative* access to well-being. If what I have said about the

---

[10] Prominent minimalist theories are elaborated by Rawls 1999 and Miller 2007, and are discussed in more detail in Armstrong 2012, chapter 3.

significance of biodiversity to human flourishing is compelling, then egalitarians should endorse the claim that it is unjust if some people can access diverse and healthy ecosystems while others cannot.[11] Egalitarians can readily embrace a negative duty not to bring about the loss of biodiversity whenever this means that people are no longer able to enjoy a vibrant and healthy environment. Such a duty would be defeasible only in cases where something of morally comparable importance was at stake—as when, for example, the basic rights of at least some, and perhaps many, people are dependent on destruction. They can also embrace a positive duty to assist people who are unable to flourish because they are exposed, through no fault of their own, to an impoverished environment. As such egalitarians can invoke the language of crisis in a more expansive way, applying it to cases where biodiversity loss undercuts people's ability to lead flourishing lives, and not only to cases where it threatens their very survival.

Such an argument will again plausibly have significant intergenerational implications. Unless we are prepared to discount the interests of future people inappropriately, egalitarians should accept that it can be unjust for us to bring about a situation in which they have access to a severely degraded environment, allowing their basic rights to be met but offering them meagre opportunities to interact with biodiversity. Although we do not know these future people or what their particular plans and projects will involve, it seems unlikely (at least in the near future) that they will differ very much from us in the interests or capabilities that matter to them. We can therefore embrace a duty of moderately strong sustainability, which among other things would require us to leave behind a bountiful and diverse living world for our descendants to engage with (see Armstrong 2021). On this kind of basis some have argued that we can have duties to future people to prevent or reduce species loss, for instance (Feinberg 2017: 375). Causing extensive species loss might irremediably impair future people's ability to engage with a diverse and vibrant living world, and if so this provides *one* reason why we ought to make serious efforts to prevent some branches of the 'evolutionary tree' being lost forever. It seems, then, that we can intelligibly speak of duties of justice to preserve diverse ecosystems, in the interests of the flourishing of both current and future people, even in cases where basic rights are not at stake. We can also intelligibly speak of duties to preserve *particular* assemblages of biodiversity for the sake of the members of specific communities.

---

[11] Egalitarian theories of global justice are discussed in Armstrong 2012, chapter 2. Examples include Beitz 1979, Tan 2004, Caney 2005, and Moellendorf 2009. Armstrong 2017a develops a global egalitarian theory of natural resource justice, which contains a discussion of conservation burdens.

Members of other species also have rights worthy of protection

The move from distinctive capacities to interests and then to rights is familiar enough in the human case. Almost all humans have the capacity to feel pain, for instance, and the great majority of them have the potential to exercise a good degree of authorship over their own lives. On the plausible assumption that these things are bad for them, we can say that people have a powerful interest in avoiding being exposed to painful experiences, or to situations in which they lack the most basic ability to determine their own fates. If so, there is a good case for endorsing a strong and basic right against torture, which is by design both extremely painful and a radical assault on a victim's agency. The same kind of movement from the identification of capacities, to the recognition of important interests, to the acceptance of specific rights is also highly plausible in the case of other living beings. This is not the place to rehearse a full list of the rights that members of each species should possess, which would be a laborious process, since the capacities typical of distinct species differ considerably. Elsewhere I have defended a set of core rights for all cetaceans, but this is just an illustration of the kind of case that would need to be made (see Armstrong 2022, chapter 7). Any rights would then plausibly be linked to duties to respect, protect, and fulfil them.

More specifically, zoocentric views suggest that the demands of justice apply to all (or at least very many) animals (see, e.g. Nussbaum 2022, Donaldson and Kymlicka 2011). Such views clearly have major implications for biodiversity conservation. To be more precise, concern for the interests of non-human animals supports biodiversity conservation in two ways. First, since biodiversity is *constituted* by individual living organisms, destroying it will typically involve harming such organisms, many of whom will be animals. This gives us a direct reason for promoting conservation. Second, concern for the interests of non-human animals gives us powerful instrumental or derivative reasons for caring about the biodiversity crisis. The *effects* of uncontrolled biodiversity loss will severely impair animals' ability to function, affecting their capacity to feed themselves, to drink, and to reproduce. Like climate change, the destruction of habitat threatens the ability of ecosystems to maintain healthy states, exposing the creatures that comprise them to malnourishment, as well as the stress incumbent on human incursions into their territories.[12] Marine animals will be threatened by the

---

[12] I use the term territories here in a descriptive sense, meaning the areas over which a particular animal or population would normally roam. For a normative defence of territorial rights for animals, see Donaldson and Kymlicka 2011, and for a critique, see Ladwig 2015.

collapse of oceanic food webs generated by ocean warming and acidification. When forests are razed, the diminution of breeding populations will lead individual animals to experience stress when attempting to find mates, and the collapse of food sources will threaten their ability to rear any offspring they do give birth to. When we come to assess our political priorities— whether to mitigate biodiversity loss gradually or rapidly, how to weigh the biodiversity crisis against other political priorities, and so on—this should be seen as an important and ineliminable dimension of that moral reckoning. In fact it is plausible that the harm that biodiversity loss will do to non-human animals provides the single most weighty reason we have for seeking to arrest or reverse it. Even if we endorse a hierarchical view of moral status— which (comparatively) discounts the interests of non-human animals simply because they *are* non-human—we must recognize that humans are *hugely* outnumbered by other animals.[13] Even if their capacity for well-being is much lower than ours, and even if we have grounds for morally discounting what well-being they can enjoy, it is still hard to believe that the harm the biodiversity crisis will do to humans outweighs the harm that it will do to the vast number of other animals we share the planet with.

My position in this book is a zoocentric individualist one, which holds that it is individual living animals that have interests (and are therefore candidates for the possession of rights), rather than inherently collective entities such as species or ecosystems. But this does not mean that we cannot possess reasons—including reasons of justice—to preserve particular species or ecosystems. In fact it seems highly likely that we do (see also Nussbaum 2022). For instance, we plausibly have powerful instrumental reasons to act so as to reduce the degradation of ecosystems, because of the ways in which the fates of so many animals are bound up in their continued health and vitality. By the same token, we will have reason to preserve species if, as ecologists suggest, species diversity is strongly linked with the broader health of ecosystems (with the implication that species extinction will be bad for many of the animals that constitute those ecosystems, even in cases where they do not belong to the relevant species). In these cases the duty we have to conserve species or ecosystems derives from duties to respect, protect, or fulfil the rights of individual animals. I will not claim, by contrast, that we can owe duties of justice to species or ecosystems per se (cf. Rolston 1985), or that they can possess

---

[13] I noted in the Introduction that humans make up perhaps a third of extant mammalian biomass. But mammals represent a tiny minority of animals, being outnumbered enormously by creatures such as fish, birds, and invertebrates. The number of marine vertebrates alone runs to many trillions. For useful data, see Brian Tomasik, 'How Many Wild Animals Are There?' https://reducing-suffering.org/how-many-wild-animals-are-there/ (accessed 25 January 2022).

entitlements in their own right. The claim that ecosystems themselves can be subjects of justice (see e.g. Schlosberg 2007, chapter 5 and Dryzek and Pickering 2019: 71) takes us onto much less certain ground. Still, the claim that non-human animals have rights worthy of protection is capable of vindicating many of the most strongly held commitments of those concerned with ecological justice. Clarifying those rights is a crucial step in understanding the nature of our duties to conserve.

Other elements of the living world possess intrinsic value

Entities that possess intrinsic value are valuable in themselves, and not only because of the goods or benefits they happen to supply to humans or indeed to other animals. There is a vigorous meta-ethical debate about whether their intrinsic value somehow 'flows from' humans, or is independent of it; but that disagreement is not important here.[14] What is important is the ethical claim that, when things have intrinsic value, we have a *prima facie* reason to conserve them (Cohen 2013, chapter 8). This reason should be seen as non-derivative in character in the sense that it is not grounded on the good that the entity in question does for anyone else. Though some have found the idea of intrinsic value mysterious, the basic idea should not be: humans themselves are widely thought to possess intrinsic value, which is why it matters in and of itself that we are able to flourish in certain ways.[15] In that sense, humans—and their flourishing—should be treated as ends in themselves, and never merely as means to others' ends. We can plausibly make the same claim about other animals too (see, e.g. Korsgaard 2018).

Indeed, our first, second, and third reasons for conserving biodiversity can all be seen to draw on the idea of intrinsic value, inasmuch as they ground duties to conserve on the non-derivative importance of protecting the interests of humans and of other animals respectively. In that sense the arguments that I have canvassed so far suggest that biodiversity has (very considerable) instrumental significance insofar as its protection advances the interests of creatures, which in turn possess intrinsic value. But it is also possible that

[14] Here, and throughout the book, I therefore use intrinsic value simply in the sense of non-instrumental value. For a useful delineation of the ways in which value might be said to be intrinsic, see O'Neill 1992.
[15] Some conservation scholars have argued that talk of intrinsic value is best avoided. For a very good critical assessment of that claim, see McShane 2007. The upshot is that the notion of intrinsic value per se is hardly something we can afford to do without. As Samuelsson (2010: 525) points out, even the most anthropocentric views must rely on the idea of intrinsic value, even if they happen to disagree with eco-centric views about the kinds of things that possess it. The view that nothing has intrinsic value, as he nicely puts it, 'is not anthropocentrism (any more than it is non-anthropocentrism); it is rather moral nihilism'.

things *other* than individual living beings possess intrinsic value, and if so this may underpin a more capacious account of our reasons for engaging in biodiversity conservation. Entities that possess intrinsic value might include plants, as well as more complex phenomena such as systems, assemblages, or relationships, or indeed features or properties of states of affairs (such as diversity or complexity).[16] That value would not reduce down without remainder to the value possessed by the individuals that make up those phenomena. It is also conceivable that, in some cases, the value in question would be at least partly dependent on relational properties such as rarity or the fact that they had not been created by human hand.

The idea that both plants and complex assemblages such as ecosystems or species possess intrinsic value is one way of making sense of the claim that they have ethical significance in their own right. I do not claim, as some ecocentric theorists have, that plants, species, or ecosystems possess rights. Neither do I claim that we possess duties towards them in their own right. However, it seems plausible nevertheless that we can have moral reasons to conserve them in virtue of their intrinsic value. The view that a unique species can possess intrinsic value in its own right, for example, helps make sense of the widespread intuition that, other things being equal, we have additional reasons to protect the last breeding population of a given species, rather than an equivalent number of (otherwise similar) individuals from a much more abundant species. When a species goes out of existence, its entire history is lost. If it is indeed true that such outcomes are in one respect worse, this suggests that species can have ethical significance even if they are not subjects of justice in their own right (Nussbaum 2006: 357).[17] Likewise, if a complex ecosystem were damaged or destroyed, we could intelligibly say that something of value had been lost even if—as seems highly unlikely—each living constituent of that ecosystem managed to find refuge elsewhere. Such a thought could again ground a reason to avoid the loss of entire ecosystems. A similar thing could be said about the loss of diversity as a property of species or ecosystems.

So far, though, I have spoken of *reasons* to conserve. To say that something has intrinsic value certainly suggests that we have a reason to preserve

---

[16] There are two ways of glossing the idea that properties such as diversity can possess intrinsic value (McShane 2017: 156). The first suggests that diversity is valuable in itself so that the more diversity is exhibited in the world, the better the world is. The second suggests that it is entities such as species or ecosystems that are valuable in themselves, but that diversity is a value-adding property for such entities. I stay neutral here on which gloss should be preferred.

[17] Note that claiming a species can have intrinsic value does not commit us to seeing it as static or homogenous, or to believing that we ought to preserve it in aspic. Like languages and cultures, species are diverse, dynamic over time, and often borrow from the outside. For a good critique of the idea of *prima facie* duties to preserve the genetic integrity of species, see Rohwer and Marris 2015.

it (Cohen 2013). It is more controversial whether the fact that an entity possesses intrinsic value suffices to ground a *duty* to preserve it, still less a duty of justice. After all, many things plausibly have intrinsic value, and it is unlikely that we can preserve all of them. We also have many other moral and political goals, which conserving things with intrinsic value must compete with for attention and resources. Regret might be an appropriate response whenever something of intrinsic value is destroyed (see Moellendorf 2022, chapter 8), but we cannot move straight from the identification of intrinsic value to a duty to preserve it in any particular case (O'Neill 1992). Indeed, there will be many cases where an appropriate degree of respect for individual animals' rights *constrains* what might be done to conserve biodiversity. For instance, captive breeding programmes that aim to secure the future of some species might be difficult or impossible to justify if they involve serious violations of the rights of individual animals, such as the right to move freely (Muka and Zarpentine 2023).

Nevertheless, it appears likely that both consequentialist and deontological theories can endorse duties to preserve biodiversity in some cases, in virtue of its intrinsic value (Elliot 1992: 148–51), at least in cases where this does not involve serious violations of individual rights. That would be a duty held in virtue of biodiversity's value, but it would most plausibly be construed as a duty owed *to* human others, and perhaps to non-human animals, too. Those others would then have a *pro tanto* right that we take appropriate action to preserve biodiversity.[18] Though *pro tanto* rights are defeasible when more important interests are at stake, if the intrinsic value of biodiversity is one factor we ought to consider when weighing up our political priorities, it then seems likely that there will be cases—perhaps many—where it makes a difference to our all-things-considered decisions about what, and whether, to conserve. Some scholars argue that appeals to the intrinsic value of collective entities or features—such as biodiversity generally, or species, or ecosystems—are best avoided, given that a focus on individuals and their intrinsic value is sufficient to motivate biodiversity conservation (Baard 2021). But even if parsimony is, other things being equal, an admirable feature of a normative theory, it must be weighed against other important features such as coherence, consistency, and completeness. If it is true that plants, species, and ecosystems have value in their own right,

---

[18] Fabre (2022) recently provided a penetrating defence of duties to conserve cultural heritage, which portrays our duties (of justice) to preserve or protect that heritage in virtue of its universal value, but where that duty is held *to* other human beings, whether past, present, or future. Though cultural heritage no doubt raises somewhat distinct questions, I believe her account of the move from intrinsic value to duties of justice is plausible in the context of biodiversity conservation as well.

then this ought to make a difference to conservation practice. A conservationism that makes no reference to them will be a relatively impoverished one, and it will (at least sometimes) make the wrong choices.

Conservation beyond biodiversity

The biodiversity crisis merits very serious attention on par with other massive global challenges such as climate change—and grappling with it requires us to engage in turn with important questions about global justice. But even when it comes to the conservation of the living world, biodiversity is far from the *only* thing we have reason to conserve. This should already be obvious: I have argued that individual animals have interests, and those interests would be worthy of protection even if we could imagine cases where doing so made little or no difference to biodiversity. One of our major reasons for conserving biodiversity is precisely that doing so helps to protect those interests; those interests have independent moral force (just as I suggest that biodiversity itself can have value independent of whether it serves individual animal or human interests). This indicates that biodiversity should be but one, albeit very important, concern of biological conservation. Although it has come very much to the fore within global conservation politics—enjoying pride of place, for instance, within the Convention on Biological Diversity—it is important that it does not crowd out other important conservation objectives. It is also important that it is not conflated with them. In focus groups with members of the Scottish public, Anke Fischer and Juliette Young found that it was common to define biodiversity as a synonym for 'wildness', for 'local species', or simply as 'a fancy word for nature' (Fischer and Young 2007: 274).

But biodiversity is not all of those things, and it is not all that matters. Although I focus on biodiversity in this book, in this section I briefly discuss some other qualities of the living world that can plausibly matter to the conservationist.

Abundance

Imagine an ecosystem comprising a great many species, each of which has suffered a considerable reduction in numbers because of human predation of one kind or another. Conceivably, human ingenuity could see an effective rear-guard action against further species loss, even as the ecosystem as a whole continued to diminish in biomass or in the number of individuals.

At the limit, we might be left with a kind of Noah's Ark version of the original ecosystem, with each population reduced down to the minimum number capable of breeding effectively. Would we have achieved everything that the conservationist should want to achieve? The answer is no. As well as maintaining variation, sustaining and/or recovering the *abundance* of the living world is another important objective.[19]

Sometimes, this point is accommodated by integrating abundance *within* the goal of preserving biodiversity. According to the authors of a new Multidimensional Biodiversity Index, for example, the 'biodiversity state' of a country ought to be measured according to three dimensions: diversity, abundance, and function (Soto-Navarro et al. 2021: 936). A well-known international database on biodiversity loss, meanwhile, defines biodiversity in terms of both the percentage of the original number of species that remain in a given area, and their abundance there (Hudson et al. 2017).[20] But we should resist the kind of mission creep whereby biodiversity comes to stand in for the bounty or quantity of life in general. It is much more plausible to use the term biodiversity to refer to the degree of variation (albeit of various types), and to acknowledge that abundance is also valuable in its own right. Abundance certainly appears to be instrumentally valuable in a variety of ways, including ways conceptually independent of diversity. Larger or fuller ecosystems are likely to be more robust in the face of environmental pressures, even if we hold diversity constant. Populations of some predator species, meanwhile, cannot be sustained without great abundance lower down the trophic scale. The human interest in exposure to rich green spaces is likely to be best served when such spaces are both diverse *and* abundant: visiting a small city zoo, for all that it is an ecologically diverse place, is no substitute for the ability to visit large and plentiful ecosystems. All of this suggests that the overwhelming focus on diversity within the contemporary law and politics of conservation should not be allowed to crowd out other dimensions of value.

[19] The Noah's Ark example does not offer a perfect contrast between variation and abundance, because while it contains lots of species diversity, it would probably contain relatively little genetic diversity. In that sense, actions to preserve variation and abundance will to a significant extent converge in practice: a genetically diverse world is likely by definition to be an abundant one.

[20] There are many other examples. One influential early definition of biodiversity, from the US Congress's Office of Technology and Assessment, declared that 'the term encompasses different ecosystems, species, genes, and their relative abundance' (see Farnham 2017: 12). More recently, the IPBES refers to 'changes in distribution and abundance over time and space' within its official definition of biodiversity (https://ipbes.net/glossary/biodiversity).

Nature and the wild

Many debates within environmental ethics have been preoccupied with the concepts of 'nature' or 'wilderness' and their centrality to conservation. This has been especially true in the US, where the environmentalist movement has often focused quite strongly on the goal of wilderness preservation. Again, these concepts are clearly quite distinct from the idea of biodiversity: we could imagine a genetically engineered world that was highly biodiverse but that contained little or no wilderness, and where the salience of the concept of nature was also uncertain.[21] Conceptually, the idea of biodiversity does not depend on the existence of wilderness or wholly natural spaces, or indeed on any division between 'nature' and 'the human' (Maier 2013: 114). Biodiversity can not only be identified within relatively untouched places, but also within urban parks and gardens. It can sensibly include humans and any creatures we happen to have engineered, as well as places or ecosystems relatively untouched by us. One salient feature of the focus on biodiversity is that it allows us to keep open the possibility of a form of conservation in which humans and their cultures are self-consciously integrated into the (rest of the) living world, rather than separated from it.

By contrast, it is a matter of considerable controversy whether we should make nature or wilderness the (or even *a*) focus of conservation efforts. For instance, if by wilderness we mean a pristine environment untouched by human interference, it has been argued that no such thing exists—in which case it can hardly serve as the touchstone for global conservation efforts (see, e.g. Callicott and Nelson 1998, Marvier et al. 2012). That claim might set the bar too high, and there are certainly places—like the deep seabed—that have hardly even been explored by humans. But it is true that wilderness is a term we should approach with caution. As William Cronon (1996: 15–16) argues, US 'wilderness' areas were not so much found by colonizers as constructed by them, in no small part by excluding Indigenous American inhabitants who had enjoyed a long and mutually constitutive relationship with local ecosystems. Cronon goes on to suggest that seeking to export such notions of wilderness to other countries can be considered 'an unthinking and self-defeating form of cultural imperialism' (1996: 18). It has long been claimed that the focus on wilderness is an unhelpful obsession of people in wealthy countries, and one that is often foisted on global South communities (which are more used to living side by side with non-human life), even

---

[21] For a careful exploration of the conceptual differences between wilderness and biodiversity, see Sarkar 1999.

at the cost of their exclusion (Guha 1989). This is a theme we return to in Chapter 6. Another suggestion is that foregrounding 'nature' or the 'natural world' as the object of conservation efforts is perilous if it leads us to forget that humans are a part of the natural world, too, and ought to be treated as such (Vogel 2011).

However, it is not obvious that those who care about conservation should avoid any reference to the origin of biodiversity (I take it that terms like 'natural' and 'wilderness' are relational terms, denoting a relative absence of human interference). To the extent that biodiversity has intrinsic value, that value might sometimes reside at least partly in (or be conditional on) the fact that it has unfolded in its own way, relatively untouched by human hand,[22] or even in the fact that it is irreplaceable or unrestorable. This is perfectly compatible with holding that in other cases its value lies partly in the fact that it has been cultivated by humans (Kallhoff 2021), just as the intrinsic value of a painting can reside partly in the relational fact that it has been intentionally produced by a human hand rather than by an accidental spillage of paint. To the extent that practices of cultivation can be hugely valuable to people, part of the value of biodiversity to humans may sometimes lie in the fact that they have cultivated it, and sometimes lie in the fact that they have not.

All of this further reinforces the point that the undoubted importance of biodiversity should not be allowed to crowd out other potential aims of conservation projects. For instance, people engaged in conservation activities need not be dismissive of concepts such as 'rewilding', which involves action (or inaction) aimed at allowing particular ecosystems to recover in their own ways from erstwhile human impacts. If the intrinsic value of ecosystems can at least partly reside in the fact that they were not created by human hand, then the claim that the balance between human-directed and non-human-directed spaces ought to be redressed is perfectly intelligible. The focus on biodiversity does not by itself tell us whether such a goal is valuable; the desire to preserve biodiversity is not the whole of conservation. It is also perfectly possible—and in my view highly plausible—to argue that our societies ought to forge a new relationship with the rest of the living world, based on peaceful coexistence and mutual respect, rather than intervention and domination. That new relationship might represent an important way forward in a world where the environmental limits to growth are becoming ever more apparent. That too would have implications for global justice. If the environmentalist

[22] We can also imagine instrumental or relational arguments for the same conclusion. As Nicole Hassoun and David Wong (2015: 188) plausibly argue, 'To remake the world entirely in the image of humanity's purposes is to limit the possibilities for our learning from that which has not been bent to our wills.'

must be committed in some senses to cutting, avoiding, and limiting human impact, it becomes still more important that they show the way forward to new ways of life in which we can be more, rather than less, fulfilled (Eckersley 1992: 17–21) in sustainable ways, and in which the ability to lead flourishing lives is fairly shared.

Conclusion

In this chapter I clarified the nature of conservation as an activity. I also discussed reasons why biodiversity conservation should be thought of as a major concern of justice, such that we can intelligibly speak of duties to engage in or lend support to conservation activities. One straightforward reason is that human life depends upon the existence of healthy ecosystems. Even relatively austere accounts of global justice should accept that we ought to cooperate together to sustain the ecological preconditions for minimally decent human lives. Biodiversity can also be important to flourishing lives, and for this reason accounts of global justice also ought to show a concern for cases in which people's access to it is better or worse, even in situations where they can all live minimally decent lives. Further, I argued that biodiversity is important because it supports the well-being of members of many other species. The fact that non-human animals possess rights of their own does a great deal to explain why we are duty-bound to support conservation policies. Finally, I suggested that biodiversity—and components of the living world such as individual plants, species, or ecosystems—can be intrinsically valuable. It is more controversial whether and when this can lead to duties to conserve, but I argue that in at least some cases it is plausible that it can.

Someone concerned to provide a full account of ecological justice would surely go on to do much more than simply laying out our reasons for pursuing biodiversity conservation. Among other things, they might go on to establish a heuristic framework for dealing with any trade-offs between these various reasons for biodiversity conservation. But my concerns here are more modest. It is enough, for now, to show that conservation will, in many cases, plausibly be required by justice. Moreover, I have suggested that conservation should be accorded much higher political priority than it has received to date. The point is that if conservation is to be taken seriously, it will have costs. We turn now to the question of how those costs should be shared.

3
Sharing the burdens

Chapter 2 provided a number of reasons why conservation might be required by justice. In many cases we must preserve biodiversity because doing so is key to securing people's basic rights or to the preconditions for flourishing human lives. Very often we must engage in conservation because doing so is key to securing the rights of animals. And at least sometimes, we must do so because of the intrinsic value of biodiversity. I will not attempt the Herculean task of establishing exactly which bits of biodiversity ought to be protected and when. However, I have suggested that the kind of radical conservation goals that have now begun to proliferate bear some plausibility. In Chapter 6 I scrutinize Half Earth proposals, which would set aside half of the land and ocean for biodiversity. But it is not only in Half Earth circles that radical ideas can be encountered. The Kunming–Montreal Global Biodiversity Framework (GBF) of December 2022 also includes objectives of grand ambition, including the target of reducing the loss of areas of considerable biodiversity importance to 'near zero' by 2030, and the goal of 'substantially increasing the area of natural ecosystems by 2050'.[1]

The key point for our present purposes is that there is a long overdue discussion about how any costs that arise from pursuing such goals ought to be shared. First and foremost, policies that misallocate conservation burdens are intrinsically unjust insofar as they let some people evade their responsibilities while requiring others to pick them up instead. Second, even if misallocating burdens is at times compatible with good conservation outcomes, in many cases it will undermine support for, and hence compliance with, conservation policies (Dawson et al. 2021: 8); as such, even those who claim to care about effectiveness alone have instrumental reasons for taking fair burden sharing seriously.

Unfortunately, *un*fair burden sharing is very common—the rule, rather than the exception—within the global politics of conservation. Worldwide, an estimated 94 per cent of all conservation funding is spent within upper-income countries—despite the fact that approximately 60–70 per cent of

---

[1] https://www.cbd.int/article/cop15-cbd-press-release-final-19dec2022

Global Justice and the Biodiversity Crisis. Chris Armstrong, Oxford University Press.
© Chris Armstrong (2024). DOI: 10.1093/9780191888090.003.0004

the world's biodiversity is to be found in lower- and middle-income countries (Stark et al. 2021: 2). This leaves the poorest people to bear heavy
burdens —people who can ill afford to take them on (Adams et al. 2004).
Within countries, meanwhile, burdens are also pervasively misallocated,
since poorer regions tend to play host to larger amounts of biodiversity,[2]
whereas cost-sharing mechanisms are often weak or non-existent. While
schemes aiming to channel conservation funds to poorer regions of the world
do exist, they are drastically underfunded. For example, protected areas have
played an outsized role in conservation politics to date, although these are just
one element of the conservation toolkit. The Global Environment Facility
has so far supported 95 protected area projects in the global South—but one
report suggests it has spent a mere 0.2 per cent of what it would take to maintain a 'globally representative protected area system' (Waldron et al. 2020).
Protecting 30 per cent of the planet's surface for biodiversity might require a
budget of between $103 and $107 billion per year, it is estimated, compared
with a current total conservation spend of $24.3 billion annually (ibid: 2020:
5). The GBF actually calls for even more than that, suggesting a minimum
spend of $200 billion per year, albeit with a surprisingly modest $20–30 billion of that flowing in the direction of 'developing' countries.[3] These numbers
are very large, but they would not make a serious dent in the global economy.
By way of contrast, approximately $50 trillion has currently been earmarked
for infrastructure projects worldwide (Watson et al. 2021: 898). This suggests a profound misdirection of resources, given that stemming the tide of
biodiversity loss is essential for the world's economies to have any future
at all.

The systematic underfunding of conservation sometimes means that biodiversity is simply lost. But it also frequently means that locals do incur costs
which others refuse to share with them. Significantly, in many such cases
the benefits of conservation accrue much more widely (Green et al. 2018).
Ecosystems in one place produce public goods—including carbon sequestration, nutrient cycling, and protection against storm damage—which are
then enjoyed elsewhere. By contrast, the destruction of biodiversity can mean
increased exposure to flooding, wildfires and pandemics on meso- and sometimes macro scales (Waldron et al. 2020: 10). We have also seen that the
ecological demands of Northern lifestyles are often placed on biomass located

---

[2] As we noted in Chapter 1, this trend is often reversed in cities, where the affluent typically have better
access to biodiversity. See Leong et al. 2018.
[3] https://www.cbd.int/doc/c/e6d3/cd1d/daf663719a03902a9b116c34/cop-15-l-25-en.pdf

in the South, in a process sometimes called environmental load displacement (Hornborg 2006). Without this displacement of ecological demands onto biodiversity in the South, the lifestyles of the affluent would scarcely be possible. This conjunction of facts should be a starting point for any discussion of biodiversity and global justice: poor people are continually entreated to make sacrifices that will ultimately benefit people in more affluent countries; these sacrifices hugely outstrip the relatively modest flow of conservation funds from North to South, for which people in the global South are presumably meant to feel grateful. In this chapter I suggest that the citizens of the global North are to a great extent guilty of free-riding on the South by refusing to pay their fair share of the costs of biodiversity conservation even in cases where they clearly benefit from it (see also Armstrong 2016). In Chapter 4, I argue that many real-world conservation projects are exploitative, in the sense that they take advantage of the weak structural position of people in the South in order to deny them opportunities they are entitled to.

Of course, none of this is to suggest that biodiversity conservation is an exclusive concern of people in the North. It is true that formal conservation organizations are disproportionately located in Northern countries; it is also true that they often have a relatively affluent membership even by the standards of the North. But conservation politics has also been marked by a powerful, if sometimes less formal, 'environmentalism of the poor', with people in the global South fighting bravely to secure the ecological basis of their livelihoods in the face of extractivism and dispossession (Martin 2017: 2). By contrast to many conservation organizations, in the United States in particular, the environmentalism of the poor is typically focused not on goals such as preserving pristine wilderness, but on securing 'sheer survival' in the face of environmental degradation, and on questions of social equity and political inclusion within conservation politics (Guha 1989). The key point here is simply that, to the extent that conservation is a global concern, this misallocation of burdens represents a considerable injustice, sustaining tragic choices between conservation and the escape from poverty when fairer options ought to be open to us. The Convention on Biological Diversity has long emphasized the importance of burden sharing.[4] But it has left the question of who ought to bear which burdens, and why, surprisingly open (Pickering 2022: 167). This chapter addresses that question directly.

---

[4] United Nations Convention on Biological Diversity (1992), Article 20.2.

The sheer diversity of conservation burdens

Intelligent policymaking requires a recognition that conservation burdens can vary in many ways: in terms of the types of activity that give rise to them; in the form burdens take; and in the actors they are borne by. Consider types of activity first. Chapter 2 distinguished between three kinds of conservation activity: avoidance, protection, and restoration. Each of these will routinely generate burdens for some actors. People aiming to protect or restore biodiversity may need to pay salaries, buy vehicles and computers, conduct surveys, and lease land. In the empirical literature on protected areas, these are often called 'management costs' (see, e.g. Green et al. 2018). Protection and restoration can also generate what practitioners call 'damage costs'. Consider for instance the destruction that can occur around protected areas when animals trample or eat crops, or flatten homes and powerlines (Naidoo et al. 2006). One study has found that in the areas surrounding a protected area in East Africa, between 4 and 9 per cent of maize yields were lost to 'wildlife damage' (Green et al. 2018: 4), and in such cases the poorest are usually found to be most vulnerable (Martin, Akol. and Gross-Camp 2015: 174). At the extreme, protected animals can threaten human life. Human–elephant conflict in Sri Lanka, for example, is estimated to cause the death of 50–60 humans per year, as well as 225 elephants (De Silva and Srinivasan 2019: 184). People who commit to avoidance, meanwhile, will incur 'opportunity costs'. As an initial gloss, these correspond to the notional benefits that would otherwise have accrued if a place had been used for some 'productive' purpose instead of being conserved. Opportunity costs can be massive in scale; according to one study, they are likely to represent two thirds or more of the total costs of conservation (Green et al. 2018). But I show in Chapter 4 that they are typically measured according to inappropriate baselines. In practice, opportunity costs are systematically underestimated, leaving the door open to conservation practices that are—frankly—exploitative.

As these examples help reveal, the form costs take can also vary widely. There is a widespread tendency in the conservation literature to depict conservation burdens in the language of dollars and cents. But while some burdens are financial in nature, others are not. As with the example of human–elephant conflict, conservation can sometimes lead to death, and in other cases it can set back human health. Some projects also thwart important life plans and identities; they can cause people to be exiled from particular places, and even when allowed to remain, people's sense of place can be undermined as activities to which they have long been committed are newly forbidden. Often it might make sense to call these latter costs 'cultural' burdens (Tan 2021: 4), but sometimes they go even deeper than that. According to Tuck and Yang (2012: 5), for instance, 'the disruption of Indigenous relationships to land represents a profound epistemic, ontological, cosmological violence' with enduring consequences. Not only cultural traditions, but also one's very sense of what world one inhabits, and what one's role is in that world, can be thwarted. In some cases, it might not be wholly inappropriate to convert costs into the language of dollars and cents, but in other cases it clearly will. The key point is that the difficulty of reckoning and comparing costs should not be taken as license to disregard some of them in conservation policy making. Refusing to even acknowledge what cannot easily be counted would be a serious flaw in any conservation policy.

Finally, it is also important to bear in mind that conservation burdens could in principle be distributed among a whole variety of different actors. They can also be allocated at a variety of institutional and geographical settings, across micro-, meso-, and macro scales. This is important because policy discussions have often been excessively focused on 'area-based' conservation measures, including, most famously, the maintenance of protected areas. In practice, this means attention centres on the (purported) need to preclude various human activities in the immediate environs where biodiversity is to be found. I argue in Chapter 6 that this focus on protected areas should be reduced, both in favour of other forms of area-based conservation that are more compatible with social justice, and in favour of non-area-based conservation policies that address a much wider range of drivers of biodiversity loss.

Protected areas are certainly *one* tool for securing biodiversity conservation, then, and they clearly raise important questions about justice. For instance, if protected areas are established in the global South, should management, damage, and opportunity costs be borne by locals, or pooled at the global level? Who should be involved in designing and managing protected areas? And who has a right to determine where these areas are located? Nevertheless, it would be short sighted and politically regressive to allow the discussion about protected areas to dominate discussions of conservation policy. For while protected areas can deal with *some* drivers of biodiversity loss (when these are linked to the actions of local communities), there are other drivers they do not engage with. In Chapter 1, I noted that the focus on the most proximate drivers of biodiversity loss has often led to the embrace of policies that coerce people in the global South, while deflecting attention away from wider structural processes that shape and incentivize their practices (Pascual et al. 2021: 569), or the environmental impact that flows from the consumption habits of the affluent (see also Duffy et al. 2019). Such an approach will be unfair if it places the onus on the disadvantaged to change their practices while exonerating privileged people who may subsidize, incentivize, and reap most of the rewards of those practices. And it will be ineffective if it leaves intact the deeper and more structural drivers of biodiversity loss.

Any progressive discussion about conservation burden sharing has to keep the unsustainable and inequitable practices of people in the global North firmly in view. As noted in Chapter 1, inequality itself is a powerful driver of biodiversity loss. An important step is therefore to recognize that efforts to constrain, or redirect, the consumption habits of the privileged can intelligibly be called conservation policies, and ought to be squarely on the political menu. Examples might include prohibitions on the consumption or import of some goods that are intractably associated with environmental damage. But in general, it probably makes sense to focus less on individual consumption decisions, and more on shifting the context and incentive structure within which people make those decisions. The withdrawal of environmentally damaging subsidies would be a good example of a progressive conservation policy, as would the intelligent use of tax policy to promote less harmful forms of consumption. Given its massive contribution to biodiversity loss, both policies could profitably be employed to reduce meat consumption, where alternatives are available (Sebo 2022: 96). Since climate change is a major driver of biodiversity loss, pro-conservation policies could also include support for the transition away from fossil fuels and towards renewable forms of energy, with fossil fuel subsidies being withdrawn and green technologies promoted. In the case of the ocean, establishing marine protected areas—which forbid or downscale at least some extractive activities within specified boundaries—would be one way of protecting biodiversity. But alternatives include efforts to reduce carbon emissions, laws that limit nitrogen pollution arising from agriculture, and the withdrawal of perverse fuel subsidies for industrial fishing.

Some would also consider measures to reduce human births an important political lever (e.g. Cafaro et al. 2022), given the relationship between biodiversity loss and the rising human population (e.g. McKee et al. 2004). Nevertheless, I am sceptical that policies which coercively prevent people bringing children into the world are justifiable. First of all, it is hard to see what coercive policies along those lines really have to contribute. The number of human births is already declining year on year, on all continents. As a result, countries like China now have virtually no population growth.[5]

[5] https://ourworldindata.org/world-population-growth (accessed 26 February 2023).

At this stage, residual global population growth is primarily driven by increased longevity and declining infant mortality, which indicates that the focus on reproductive decisions is something of a red herring. Policies that support the education and empowerment of potential parents—and especially women—are easy to defend on grounds of freedom and autonomy without even considering questions of biodiversity. So too are policies aimed at tackling poverty, which might also lead people to freely choose to have fewer children. But neither kind of policy would coerce potential parents, and each would expand, rather than restrict, individual freedom. Second, coercive conservation policies are hard to justify if there are ways in which people can reduce their environmental impact that do not involve constraints on their reproductive choices (Caney 2020). And it does seem likely that policies aimed at reducing the material throughput of our societies (including the policies discussed in the last paragraph) have considerable potential. Third, focusing on reproduction rather than consumption looks to be politically regressive, directing our attention away from the global North (where people consume very heavily, but where birth rates have declined to an historically low level), and towards the continent of Africa (where people consume much more modestly, but where birth rates, though declining, are significantly higher than elsewhere). According to Schaffartzik et al. (2019: 72), half of the world's resources are consumed by the wealthiest fifth of the world's inhabitants. Coercive population policies therefore risk unfairness towards people in the global South, if they bear relatively little moral responsibility for environmental damage—and they may also risk unfairness towards people in the global North, if there are other ways they could reduce their environmental impacts. In sum—though all of the processes that drive biodiversity loss ought in principle to be up for discussion—it will usually be both fairer and more effective to target consumption practices, especially in the global North.

How should conservation burdens be allocated?

The recent GBF makes clear that securing biodiversity will require much more money and effort than conservation policies have attracted to date. I have suggested that conservation burdens can comprise far more than a lack of money, since they also include setbacks to culture, identity, health, and much else besides. But on what basis should any burdens be distributed? What counts as a fair, or an unfair, allocation of burdens? The answer cannot be that burdens should simply fall on the shoulders of locals, even if that is often the depressing reality. Consider again this conjunction of significant background facts: biodiversity is concentrated in the global South; so is poverty. Many threats to biodiversity arise beyond the local level; many cases of successful conservation will go on to benefit outsiders as well as insiders. There may be *some* cases where requiring locals to pay for conservation—simply because they *are* locals—is unobjectionable. But I believe that these cases will be limited to instances where conservation is purely optional—where locals freely choose to pursue conservation goals above and beyond those required by justice. In cases where conservation is *required*, it is much less clear why mere proximity matters.

It is also important to recognize that policy makers will confront many cases where there *are* no locals to pick up the tab. The continent of Antarctica is home to a surprising amount of biodiversity, including seals, birds, plants, lichen, invertebrates, and much else besides (Wauchope, Shaw, and Terauds 2019). But very few people live in Antarctica, and it is not officially part of the territory of any state.[6] If the few people who work or visit there cause environmental damage, it may well be appropriate for them to bear the costs of clean-up. But for the most part locals are not the problem: the environmental threats faced by the Seventh Continent are largely exogenous, with climate change representing the biggest threat by far. An equally difficult case is the High Seas, which represents approximately half of the surface area of the earth, and a very large majority of its habitable space. The High Seas, though, are beyond any state's jurisdiction, and the number of people who live there permanently is vanishingly small. And fish, and many other marine animals, of course, move—both throughout the High Seas, and in and out of individual states' marine territories. The treaty on Biodiversity Beyond National Jurisdiction agreed in March 2023 offers hope of modest progress in tackling the environmental problems of the open ocean (Armstrong 2023). But according to what normative principles should any burdens be allocated?

My view is that when conservation is required by justice, burdens should in the first instance be allocated on the basis of contribution to the problem, where appropriate. When that is not appropriate, policymakers should turn to capacity to bear burdens. This division of labour is familiar from debates on climate justice, and I believe it serves us well when we think about biodiversity conservation too. At the end of this chapter, though, I also consider some arguments that seek to link the allocation of burdens to the question of who benefits from conservation. Although patterns of benefits are not as important as contribution or capacity, I suggest that they can inform the way we should appraise actors' failure to accept their proper share of conservation burdens. In short, the fact that someone benefits from conservation but

---

[6] Territorial claims, while not withdrawn, were 'frozen' under the Antarctic Treaty of 1959.

wrongly refuses to share in the burdens it produces means that we may be entitled to criticize them for free riding. This, in fact, is likely a term that can legitimately be applied to many people who benefit from conservation, especially in the global North.

Contribution to the problem

If the reason a conservation-worthy element of biodiversity is under threat is that it is being degraded or destroyed by someone, then an immediately attractive solution is to ask that person, or group of people, to pick up the tab for remedying this threat. The 'Contributor Pays Principle', or CPP, would place costs on the shoulders of whoever is morally responsible for damage to conservation-worthy entities. Roughly speaking, the argument runs along the lines of 'you broke it, you fix it'. The general principle is very familiar from discussions of climate justice, where it is frequently argued that heavy emitters, for instance, ought to bear the lion's share of the costs of mitigation, adaptation, and repair (Shue 2014). More broadly, we often hold people remedially responsible for rectifying problems when they have played a significant causal role in bringing them about (Miller 2007). If someone overturns a shelf full of porcelain in a souvenir shop, the first question the shopkeeper should ask is not which customer has the fattest wallet, but rather who *did it*. The CPP taps into the idea that humans are moral agents, capable of making choices, and that—at least when certain conditions are in place—it makes moral sense to hold them responsible for those choices. The principle, of course, requires further specification. As an initial gloss, it seems to make moral sense to hold people responsible *when* they ought to understand the possible adverse consequences of their actions *and* can act otherwise without facing unreasonable costs. Where those conditions are met, there are moral grounds for placing greater burdens on the shoulders of those who have generated threats to biodiversity. For instance, if the reason a particularly diverse ecosystem is threatened is that somebody has recklessly and unnecessarily used a dangerous pesticide nearby, the CPP might hold them responsible for any clean-up costs.

That is a simple interactional case; in principle, the CPP can also be applied in more complex or structural cases, where conservation problems arise through the actions of very many people in combination. Consider climate change as an important driver of biodiversity loss. Climate change arises from the activities of billions of people—that is, the actions of an individual are not sufficient to precipitate a crisis. This, the unpredictability of weather patterns, and the fact that climatic change is likely to involve unexpected tipping points, make it practically impossible to trace the impact of any one person's emissions on biodiversity. However, this does not let us off the moral hook. We may act wrongly even when our actions have only a small chance of contributing to serious harms. In such circumstances we *might* end up being the one to make a difference (Cripps 2022: 166) but we are unlikely to know one way or the other. How should we proceed in cases where our causal contributions are epistemically opaque—that is, where it is unclear who actually made the decisive difference in bringing about significant damage to biodiversity? Barry and Øverland (2015: 183) plausibly suggest that in such opaque cases all actors who behave in the relevant way should pick up remedial responsibilities, and that the higher the chance that any particular actor made a difference, the higher the share of burdens they should accept.

In principle, groups might count among the right kinds of actors to bear contributory responsibility for biodiversity loss. But which groups? Stephanie Collins has usefully distinguished between combinations, coalitions, and collectives, all conceived of as types of group. Mere combinations of people share neither goals nor decision-making mechanisms; coalitions share goals but not mechanisms. Neither type of group is the right kind of entity, she suggests, to bear group duties. By contrast, *collectives* possess decision-making mechanisms that make the attribution of responsibility potentially meaningful (Collins 2019, chapter 1). To be clear, the question Collins is pursuing is when groups can possess 'prospective' or forward-looking moral duties to bring about certain good outcomes (ibid: 7). But one basis on which we might believe groups hold such duties is that they are retrospectively responsible for creating problems like biodiversity loss. The CPP seeks to make such a bridge between the two things and suggests that when actors bear moral responsibility for bringing about negative outcomes, they can therefore pick up 'remedial' responsibilities to put them right (Miller 2007). The way is then open to argue that collectives such as corporations and states can bear contributory responsibility for environmental harms (Tan 2023: 3), provided the appropriate decision-making mechanisms are in place.

On the other hand, we will confront many cases where the CPP appears out of place. First, there will be instances where the reason conservation is required has nothing to do with human action. If the cause of damage to biodiversity is not anthropogenic, we will need to turn to some other principle for guidance (and if the damage is only partly anthropogenic, the principle will only give limited guidance when we address our question about conservation burdens). Second, damage may have been caused by people who are no longer alive. If so, we cannot now make past generations pay the price of putting it right. We might, of course, hold their present-day descendants responsible. However, if we do so, it will not be because the latter have *contributed* to the damage done by their ancestors: the argument for holding them remedially responsible would have to rely on some other principle.

Finally, there will be cases where the people who caused damage are still living but nevertheless should not be held remedially responsible. One reason might be that they were excusably ignorant of the likely consequences of their actions. In general, we do not hold people morally responsible for harms they did not know they were causing, unless we are confident that they really *ought* to have known what they were doing. In some cases, ignorance is negligent or even wilful, and people can rightly be held to account for it. In other cases, it is more innocent, and when this is true it seems wrong to force people to make amends for harm they did not know they were bringing about. Some, certainly, doubt that we should restrict the scope of the CPP in this way. Kok-Chor Tan, for example, argues that people can have a duty to repair harms they have done, even if they were excusably ignorant of the damage they were causing. This brings us to something like a 'strict liability' account of the CPP, at least where ignorance is concerned (Tan 2023: 4). However, I believe we should be cautious about such arguments. It is often vital that people act carefully and avoid imposing serious risks on others. The prospect of serious harms to the biosphere means we can no doubt supply many relevant cases in the conservation context. In such circumstances, education about the consequences of people's actions is important. At this point the number of people excusably ignorant of the consequences of their actions on the climate is tending towards a vanishing point, and we must hope that the same can soon be said about the drivers of biodiversity loss. But the point of education is to ensure that ignorance is not excusable, so that people can then be held responsible going forward. It is less clear that genuinely innocent harms should be laid at people's doors if that can reasonably be avoided.[7]

---

[7] In the context of climate change, Simon Caney (2010: 210) suggests that greenhouse gas emitters 'should bear the costs of their actions even if they were excusably ignorant of the effects of their actions *if they have benefited from those harmful activities*' (italics in original). Accepting this argument would also bring us closer to a strict liability version of the CPP. However, I believe we also have grounds for suspicion about this suggestion. The central problem is that in the cases Caney has in mind, people do not act wrongly: it is neither the case that they understand the consequences of their actions, nor that they *ought* to do so. If so, it is not obvious why they in particular should bear costs (Wündisch 2017: 840). Although Caney's suggestion seeks to modify and extend the CPP, some have thought that these gaps should instead be filled by a separate Beneficiary Pays Principle (BPP). García-Portela (2023) plausibly argues, however, that this move does not help, because the BPP ends up facing much the same problems in cases of excusable ignorance.

The CPP will also fall short in situations where people have no reasonable alternatives to acting in the way they do, where 'reasonable' needs to be fleshed out with reference to some kind of baseline. People who are dependent for their survival on environmentally destructive activities should not normally be asked to desist from them until alternatives are offered,[8] and they cannot fairly be asked to bear the costs of restoration and repair if doing so would throw them back into penury. To the contrary, a concern for conservation will probably be best advanced by providing them with the resources to pursue alternative forms of livelihood as a matter of urgency. People who *can* bear contribution-based burdens and still remain above the average level of well-being (or, better, the average sustainable level of well-being) should bear them in full. Between those two poles, we should operate a taper so that people who would be left just above the poverty threshold bear a very small amount of contribution-based burdens, people who would be left just below an average standard of living bear them almost in full, and so on. The threshold and the taper both reflect the judgement that the contributory responsibility people bear can be attenuated in light of their circumstances. The least advantaged should not bear any costs at all, where this is avoidable, and those whose choices are relatively straitened should bear a proportionately smaller share of remedial burdens.

Capacity to bear burdens

In cases where the CPP fails, burdens should be allocated in line with the Ability to Pay Principle (APP), according to which people should bear burdens in line with their capacity to do so (Shue 2014). There is much more to be said about what constitutes capacity, of course. In my view, it should refer to people's access to well-being, but in principle we could also construe capacity in terms of the resources people have at their command. And in practice, we might need to turn to some relatively rough-and-ready proxy. Globally, for instance, an index that factors in levels of 'human development' could be used to identify the individuals or countries most able to bear conservation burdens (see Moellendorf 2014 for a parallel proposal when it comes to climate change). Either way, the principle will be attractive to those who care about equality because it asks more of those who are doing well and

---

[8] That is, they should not be asked to sacrifice their basic rights in the interests of conservation *unless* this is practically unavoidable, or all other options violate the basic rights of even more people.

avoids putting still greater burdens on the shoulders of those who are faring badly.

Consider a real-world example. Gorillas fall into Eastern and Western species, divided geographically by the Congo River, and the Eastern species is further divided into Mountain and Lowland sub-species. Mountain gorillas have by now been reduced to two populations spanning the Democratic Republic of the Congo (DRC), Rwanda, and Uganda (Granjon et al. 2020). Conservationists estimate that perhaps a little over a thousand individuals remain. The majority of them live in the DRC, a war-torn society described by the United Nations as one of the world's least-developed countries, with a shockingly low annual per capita income of US$545 in 2019 (Rwanda, in the same year, had an annual per-capita income of $820, and Uganda $737).[9] Assume what I will not seek to prove here, which is that conserving the remaining gorillas is a demand of justice. The point is that, if Mountain gorillas are to be protected and populations even restored, there will be significant costs attached. People who kill gorillas for meat or for the illegal wildlife trade must be prevented from doing so and must be given alternative ways of making a living. Habitat loss must be slowed or even reversed, which also means that those who would otherwise benefit from deforestation must be given alternative options. In short, conservation projects must reckon with local people's struggles to escape from desperate poverty. To localize burdens, in this case, is to say to the Congolese people: if you want to conserve gorillas, that is none of our concern. You must meet the costs alone. But the suggestion that some of the poorest people in the world ought to pick up the tab here seems to me to be a non-starter. In this situation, people probably bear little or no contributory responsibility: that is, they probably have few (or no) good options that do not harm gorillas or damage their habitat. In such cases, we must turn instead to capacity. If so, the right conclusion seems to be that outsiders have a duty of justice to bear conservation burdens because locals simply cannot be expected to.

The APP is usually thought to square well with egalitarian views about justice, because the poor are absolved of bearing any burdens, while above that threshold burdens rise in line with overall access to well-being. I distinguish this position from a subtly different version of the APP, which is defended by David Miller (2009), who argues that policymakers should excuse those who are poor, or very poor, from carrying any burdens when

---

[9] https://data.un.org/Data.aspx?d=SNAAMA&f=grID%3A101%3BcurrID%3AUSD%3BpcFlag%3A1 (accessed 14 December 2021). These figures are shockingly low even if we believe, as I do, that GDP per capita is generally a poor proxy for well-being. See Armstrong 2019a.

we come to deal with shared problems such as climate change (and, per-haps, the biodiversity crisis), and then require people above that threshold to bear them equally. However, this variant on the APP is objectionable: it is wrong to require people who are *very* well-off, and people who are just above the relevant poverty line, to bear equal burdens (Armstrong 2017b). The reason so many states commit themselves to (notionally) proportional or progressive tax regimes is presumably that capacity to bear burdens mat-ters not only below the threshold of poverty, but also above it. Progressive policies require people who are doing very well to contribute much more than those who are doing averagely well to the provision of public goods and other services, and they plausibly require the badly off to contribute nothing at all. If capacity matters, in short, it is implausible that it only mat-ters at or below the threshold of poverty. Our approach to capacity should be robustly egalitarian, and consistently sensitive to differences in access to well-being.

The optimal solution looks to me to be something like this. People below the threshold of poverty ought to bear no conservation burdens at all, if that can reasonably be avoided (unless all other options undermine people's basic rights even more seriously). Above that threshold, people's capacity-based burdens should rise progressively in line with their *ex post* level of well-being (their well-being after burdens are absorbed). In practice, this rendering of the APP would plausibly exempt the world's poor (who are dis-proportionately, if not exclusively, located in the global South) from bearing any conservation burdens at all, and it would place much larger burdens on the shoulders of the top 10 per cent (who are disproportionately, although far from exclusively, located in the global North).

Overall, then, my suggestion is that we should turn to facts about con-tribution and capacity when we need to allocate the costs of responding to important collective problems like the biodiversity crisis. When actors are causally connected in the right way to biodiversity threats, they ought to pick up the tab for tackling them. The idea of being causally connected *in the right way* draws our attention to the fact that there will be many cases in which actors cannot (or should not) be asked to pick up the tab, for example, because they are no longer living, were excusably ignorant of the consequences of their actions, or on grounds of capacity cannot be expected to act otherwise. In such cases, we should turn to the APP to take up the slack. The idea that we turn first to contribution and, where contribution fails, turn to capacity, enjoys some support in discussions of global justice and cli-mate change (e.g. Caney 2010, Miller 2009, Cripps 2022), and I suggest it also should be applied to the costs of biodiversity conservation.

Facts about benefits

So far, I have said relatively little about benefits. But according to one influential argument in the conservation literature, it is in fact those who benefit from biodiversity conservation who should pay for it (Balmford and Whitten 2003). This appears to be a very strong view: facts about contribution and capacity, on this argument, do not seem to be normatively relevant at all. But it is an idea with some practical resonance. The idea that people should pay when they receive benefits from conservation activities appears to be consonant with the Payments for Ecosystem Services (PES) approach, according to which those who consume environmental 'services' ought to pay people who make those services possible, for instance, by retaining or enhancing forests or other ecosystems. Globally, the PES approach is now one of the most significant market-based conservation systems, accounting for an estimated $36–42 billion in annual transactions (Salzman 2018).

However, this alternative principle will often lead us astray. First of all, requiring those who benefit from conservation to bear its costs can lead us to wrongly neglect facts about contribution to the problem. For instance, take an important but rather neglected conservation issue: the contemporary sargassum problem. Sargassum is a brown algae typical of the Sargasso Sea, where it provides a vital habitat for eels to spawn. In recent years—likely as a consequence of both climatic changes, and run-off caused by deforestation in the Amazon—sargassum beds have begun to break up, and sargassum now washes up, and eventually rots, in vast quantities on the coastlines of West Africa and the Caribbean. There it poses major health problems for locals, as well as undermining domestic fishing and tourist industries (Devault et al. 2021). If the sargassum problem was tackled at its roots—say, by reducing climate change and deforestation in Latin America—it would be coastal inhabitants in the Caribbean and West Africa who stood to benefit. But it is far from clear that they should pay for the relevant policies to be implemented, when they bear little contributory responsibility for the existence of the problem.

Second, making the beneficiaries of conservation pay wrongly neglects facts about capacity to bear burdens. That is a further problem with the sargassum scenario: since communities in West Africa can hardly afford to solve the sargassum problem for themselves, it would be unfair to expect them to do so. Sometimes the reason we ought to conserve is indeed that other people will thereby benefit. But this does not mean that those other people should then pay. Better, I suggest, to turn to considerations of contribution and capacity. The attempt to link burdens with benefits presumably gains what superficial appeal it possesses from the fact that it is often people in the global North who benefit from conservation. But in many other cases poor people will be the beneficiaries, and here the argument looks less plausible.

Still, it is possible that facts about benefiting from conservation can play *some* kind of role in determining the fair allocation of burdens, even if they are not all-important. In my view, they help to specify the *kind* of wrong people commit when they refuse to bear their fair share of conservation burdens.[10] Imagine there is an ecosystem that requires protection at the bar of justice. Members of many communities stand in line to bear the relevant conservation burdens, on the basis of their capacity to do so, or their contribution to the problem. Two of those communities, however, stubbornly refuse to make any contribution. The first community does not benefit when conservation goes ahead regardless. However, the second does, and in fact, its members actually depend upon the ecosystem functions that successful conservation secures. Here it seems clear that both communities are acting unjustly, by refusing to pick up their fair share of the burdens. But only one of them is refusing to pick up the burdens of supporting activities from which they then go on to benefit.

Does this make a moral difference? I believe it does. The first community is simply shirking its responsibilities, which is bad enough. But the second community is also free riding on others' efforts. Shirking burdens where the activity in question is in *our* interest means we commit not just the general sin of unfairly shifting burdens onto others, but the more specific sin of free riding on others' contributions. It could be that we cannot avoid benefiting from conservation, for example, because there is no other way of sustaining life in our communities.[11] It might be that we would prefer to avoid being dependent in this way. Still, that we *are* dependent, that costs need to be borne by someone, and that we are properly selected to bear some of those costs (on the basis of capacity or contribution) but refuse to do so is enough to identify us as free riders. The fact that the second community will benefit from conservation should be seen as a morally aggravating factor when we come to appraise its members' refusal to contribute. It means, I think, that we should criticize its refusal more strongly still.

---

[10] I suggested in earlier work (Armstrong 2019b) that benefiting from conservation can be relevant in filling the gaps left by the CPP, without quite spelling out how. The current discussion represents my attempt to resolve this issue.

[11] Some have thought that it is inappropriate to label people as free riders if they are not free to accept or reject the goods they go on to receive. I agree with Klosko (1987) that the charge of free riding remains appropriate in cases where they depend upon those goods. I believe what is crucial to the charge of free riding is not our attitude towards the goods we receive, but our dependence on them, alongside our refusal to bear any share of their costs. It is possible to be mistaken about that dependence, of course, which means in principle that some actors may *inadvertently* free ride on the sacrifices of others.

So far, so hypothetical. It seems to me that the issue has real-world import insofar as it captures the situation in which advantaged people in our world often find themselves when it comes to conservation burdens. People in the global North, as well as the better-off inhabitants of the South, should often shoulder conservation burdens because of the role they play in causing threats to biodiversity. Refusing to do so constitutes an injustice. Even when they do not cause threats, they should often shoulder burdens because they *can* do so at relatively little inconvenience to themselves. Again, refusing to do so constitutes an injustice. In cases where they ought to bear burdens but refuse to do so, the fact that they will benefit when conservation eventually comes to pass—the fact that their livelihoods are partly secured by others' sacrifices—makes their moral situation still worse, changing them from mere shirkers (though that is bad in itself) into being free riders, too.

Conclusion

The chapter has covered a lot of ground. It is time to draw the strands together. I suggest that, in cases where conservation is required by justice, conservation burdens should be allocated on the basis of contribution and capacity. When it comes to contribution, we should hold people remedially responsible when they knew (or should have known) the likely consequences of their actions, and where they had reasonable alternatives. In fleshing out the point about alternatives, I suggest those who would be left in poverty should not be asked to pick up any capacity-based burdens, that those who would be left above an equal sustainable standard of living should bear full remedial burdens, and that we should apply a taper in between, so that people bear a greater share of remedial burdens the closer they are to an equal sustainable standard of living.

In cases where the CPP is exhausted, we are left with capacity. Again, those who would be left below a poverty threshold should not be required to bear any capacity-based burdens. Those who would be left above that threshold, after bearing burdens, should bear those burdens in a progressive fashion, with those enjoying higher *ex post* levels of well-being bearing greater costs. Finally, I considered some ways in which patterns of benefits might be said to matter. For the most part, I argue that patterns of benefits are not especially relevant to the distribution of conservation burdens. But I suggest that, in cases where someone already stands in line to bear some conservation burdens, the fact that they depend on the benefits arising from conservation should be seen as an aggravating factor when we come to appraise their refusal to pay up. Such a refusal can mean that one is not only a mere shirker, but also a free rider. This suggestion is not of merely philosophical interest: it appears to reflect the moral situation of many of us in the global North, who neglect to (adequately) share in the costs of protecting vital elements of biodiversity, even when conservation is undoubtedly in our interests.

4
Opportunity costs and global justice

Opportunity costs can represent a significant portion of the costs associated with conservation projects (Green et al. 2018), frequently outstripping other kinds of cost (Balmford and Whitten 2003). They are typically understood to refer to the benefits someone *could* or *would* have obtained if conservation projects had not required them to give up on current activities such as farming or hunting in a particular place (Naidoo and Adamowicz 2006, Adams, Pressey, and Naidoo 2010). As Green et al. (2018: 2) put it, to identify opportunity costs we simply need to measure 'the net benefits obtained if the land were available instead for development to some other productive use'. The same can presumably be said at sea. In this chapter I show that this familiar way of identifying opportunity costs is flawed, however, and that when used to calculate what people affected by conservation projects are owed, it opens the door to considerable injustice. I show that the analysis of opportunity costs provides a good example of the importance of considering biodiversity conservation and global justice alongside one another, rather than thinking about them in isolation. I distinguish four ways of identifying opportunity costs and make the case for an egalitarian baseline. Such a baseline would suggest that, in practice, many of the world's poor are being unjustly treated, or even exploited, as a result of conservation activities.

A moralized baseline for opportunity costs

In many cases, conservation projects require people to give up on valuable economic opportunities. In such cases, they are said to incur an opportunity cost. To measure those costs, we must judge them in relation to some baseline, such as the income they *would have* earned if they had been able to perform some activity or another. Policy makers then face the empirical challenge of measuring how far conservation might cause someone to fall below the baseline in question. They would then be in a position to offset those opportunity costs, if appropriate. But before they get there, they face the *moral* challenge

Global Justice and the Biodiversity Crisis. Chris Armstrong, Oxford University Press.
© Chris Armstrong (2024). DOI: 10.1093/9780191888090.003.0005

of specifying which baseline is the correct one to use. Policy makers' answers to that question will have enormous implications for global justice.

It might be suggested that there is no great moral mystery here: to calculate opportunity costs, we simply need a description of the activity an actor would otherwise have engaged in, along with the benefits it would have brought them. In that sense opportunity costs might be thought to be a morally neutral category. To the contrary, I argue that identifying opportunity costs must involve reflection on the kinds of opportunities people *should* have access to. In some cases, the activities in which people are currently engaged should be forbidden. Imagine that a gang of people are engaged in growing a recreational drug that is hugely damaging to human health. This activity is highly lucrative, but a protected area is declared locally, and the gang asks conservationists to replace the income they would have received from selling the drug. That is not a request conservation policy makers should fulfil. The gang's members might still have a right to help in finding alternative livelihoods, if they currently lack for other options. But using expected drug revenues to calculate opportunity costs would be to select a morally inappropriate baseline. The same would go for a variety of clearly unjust activities, such as those involving slave labour or the exploitation of children. If an activity should not normally be seen as permissible in the first place, then it does not form a suitable baseline for identifying opportunity costs. It makes more sense to connect opportunity costs to the kinds of opportunities people ordinarily *ought* to have, were it not for the need for conservation at a particular site.

Using actual opportunities as the baseline for opportunity costs could also wrongly cause policy makers to give people much *less* than they are entitled to. Imagine that a community of farmers is cruelly exploited, for instance by being trapped in debt bondage. Their government does not step in to help, because the farmers belong to a marginalized ethnic group. As a result, their incomes are far lower than those earned by other locals and leave them vulnerable to serious malnutrition in lean years. If a protected area was declared in the locale, it would be quite wrong for its funders to give fewer resources to members of the exploited group, compared to others, even if it is the case that they *would have* earned less if they had been able to continue farming. Here, too, we can see that opportunity costs must be connected to a view about the kinds of opportunities that people *ought* to have.

Such cases show that the baseline for calculating opportunity costs must be a 'moralized' one, rather than morally neutral. It must make some reference to the kinds of opportunities that people should, and should not, have. But this is not the end of the matter, because we can imagine several different moralized baselines for opportunity costs, and policy makers will have to choose between them. In what follows I sketch and evaluate four possible baselines. The first two are very familiar from conservation practice, and do not draw any explicit connection with theories of global justice. The last two are less familiar, but they do both explicitly take considerations of global justice on board. I argue that the first two baselines ought to be rejected, and that the fourth is to be preferred to the third.

To keep things simple, I concentrate mainly on cases where conservation is clearly required, but where the actors who are being asked to change their behaviour simply cannot afford to bear any of the costs of doing so. As such, their opportunity costs must be met by others if conservation is going to take place. In some of the cases I discuss I use figures in US$. It is highly likely, as I suggested in Chapter 3, that many of the losses conservation projects cause are ill-captured by focusing on dollar income. This might be true of the 'cultural' costs incurred by people required to give up on traditional activities (Tan 2021), or of setbacks to health, or to the attachments to particular places or ecosystems that I discuss in Chapter 5. There is a strong tendency in conservation practice to disregard costs that cannot be measured in material terms (Thondhlana et al. 2020), but this is indefensible. It is vital that the full variety of costs are considered when appraising the permissibility of conservation policies. I use dollar figures at points in this chapter purely to illustrate the way that the various baselines diverge in their implications, and not by any means to suggest that income is all that matters.

The status quo baseline

In practice, opportunity costs are often calculated by establishing what people counterfactually *would have* earned, had conservation projects not taken place (Fisher et al. 2011, Green et al. 2018). Calculated in financial terms—as they so often are—this could mean the money they would have earned if they had been able to carry out those activities. Alternatively, opportunity costs might be calculated in terms of 'expected' value, where the revenues from the relevant activities are multiplied by the likelihood that they would have been brought to fruition (Naidoo and Adamowicz 2006). Either way, payments are linked to the status quo before the project took place, and specifically to the opportunities that then existed. It is important to recognize, however, that the status quo baseline will itself plausibly be a moralized one, insofar as it connects (or ought to connect) conservation payments to activities that could have legitimately taken place. The status quo baseline cannot plausibly be a purely factual one that describes whatever activities people *would have* engaged in. As the examples of growing dangerous drugs or using slave labour show, it ought to draw a connection with activities that people could *justly* have brought to fruition.

Why should policy makers adhere to the status quo baseline? There is certainly a pragmatic justification, insofar as people are likely to resist conservation projects that make them worse off; as a result, paying them what they would have earned might be necessary to secure their compliance. And there is a potential fairness justification, too: conservation activities should not make people worse off, we might say, if that can be avoided. In conservation practice, this thought is reflected in the familiar principle that conservation projects should first and foremost 'do no harm' (Sims and Alix-Garcia 2017).

The problem with the status quo baseline, however, is that the opportunities people have are often deeply unfair. Adopting a status quo baseline condones and may even reinforce that unfairness. Imagine that a global conservation organization has decided to make payments to farmers who agree to let their fields lie fallow. One farmer, in the US, could have made $10,000 if she had grown crops on her land rather than leaving it to rest. Another farmer, in Sierra Leone, could have made only $500. The status quo baseline suggests this is what each should receive. But what if the disparity in opportunities in this scenario emerges within a vastly unfair global economy, and is influenced by an historical legacy of colonialism? To accept *actual* opportunities as the relevant benchmark is to place the background context offered by the status quo beyond moral question.

The status quo baseline can produce problems of both overpaying and underpaying. In many cases it will involve overpaying, whereby decision makers send substantial funds in the direction of people who, even without such funds, would have a comfortably above-average standard of living (considered in global terms). But in many cases, it will involve underpaying, sending meagre funds in the direction of people who are already at the receiving end of various forms of injustice. In cases where people's opportunities are unjustly constrained, employing the baseline could mean avoidably leaving people in desperate and undeserved poverty. If conservation decision makers have it within their power to offer people a ladder out of unjust poverty, but choose not to, this could render them complicit in that poverty.[1] The global economy contributes to radical inequalities in access to well-being, and many people's prospects continue to be shaped by violent practices of

---

[1] For an influential contemporary account of complicity, to which I am sympathetic, see Lepora and Goodin 2013.

dispossession, slavery, and colonialism. There is no reason why conservation policy makers should accept that status quo as morally authoritative. Instead of accepting the status quo baseline, they ought to identify opportunity costs by thinking through the kinds of opportunities that people *ought* to have.

The willingness to accept baseline

In many real-world conservation projects, opportunity costs are identified by establishing affected people's willingness to accept compensation (Lindhjem and Matani 2012, Bush et al. 2013, Lennox and Armsworth 2013, Tadesse et al. 2021). This typically involves surveying those who are likely to be affected by a conservation project and asking them how much money it would take to make them indifferent about whether that project takes place or not. If someone states that they would require a payment of $800, say, before they accepted a conservation project that disrupted their livelihoods, then a payment of that magnitude might be seen to cover their opportunity costs. We might suppose that the willingness to accept baseline will deliver identical results to the status quo approach. We might even consider willingness to accept surveys as a useful method of identifying what people stand to lose compared to the status quo, rather than the basis for a distinct baseline. But in principle, the two baselines could pull in different directions. For instance, if a community genuinely had a veto over a conservation project, they might ask for an amount higher than the actual economic cost they would incur.

For an illustration of what the willingness to accept baseline could look like in practice, consider a study on forest conservation by Mahesh Poudyal and colleagues (Poudyal 2018a; see also Poudyal 2018b). Their study concerns a REDD+ pilot project in the Ankeniheny–Zahamena Corridor, a large, protected area in Eastern Madagascar. Under the REDD+ scheme, the World Bank seeks to defray the opportunity costs of forest conservation by providing alternative livelihood options (such as improved agricultural, livestock, and beekeeping projects) to locals who have been required, as a result of the declaration and expansion of the protected area, to give up traditional practices such as swidden agriculture (Poudyal et al. 2018a: 6).[2] As well as being a haven for biodiversity, Madagascar is home to the second-highest proportion of citizens classified as 'extremely poor' of any country in the world, and as a

---

[2] Swidden agriculture involves people cutting away vegetation, growing crops for a few years, and moving on, and is a fascinating and controversial topic in its own right. Though it is often described as 'slash-and-burn' agriculture, its environmental ramifications are highly contested (see e.g. Dominguez and Luoma 2020).

result the conflict between conservation and poverty is acute. Three findings from the study are of particular interest. The first is that more than 50 per cent of affected locals have received no assistance at all to date (ibid: 1). Second, the official level for assistance is set extremely low: the published plan suggested that each eligible household would receive a one-off payment of between $100 and $170 (ibid:16), even though the effects of exclusion would be significant, and would be felt for decades or more (ibid:19). Third, the baseline which Poudyal and colleagues suggest would be *more* appropriate (and more generous) is instructive in its own right. The authors conducted a choice experiment intended to elicit locals' willingness to accept financial transfers. After asking locals how much money they would accept in return for giving up swidden agriculture in the area, they arrived at a median figure of $2,375 (ibid:13). This, they claim, represents the *true* opportunity cost of conservation in the area, and it should therefore form the baseline for fair assistance, rather than the relatively modest payments made by the scheme to date (ibid:15).

Those higher figures move us closer to a just outcome, but they are still far too low. The problem with using willingness to accept to identify opportunity costs is that we have little reason to believe that what people would *accept*— if they had to choose between conserving and continuing as they are—and what they are *entitled* to are the same. A willingness to accept framework can undoubtedly provide the conservation planner with useful practical information about how much conservation she can get for her money in different parts of the world. But it is far less clear that it can tell her what people *ought* to receive when they are required to give up on activities to which they are committed. On the one hand, social scientists tell us that privileged actors, with plenty of alternatives, can leverage their position to extract conservation payments greater than they would in fact have received had they used their land to earn income in the formal economy (Lennox and Armsworth 2013). On the other hand, disadvantaged actors may, when giving responses to a willingness to pay experiment, 'settle' for what they expect they would have received without the policy intervention, which might be much less than they *ought* to have received, if their opportunities were fair. In the formal economy, people often after all accept exploitative wages that are far less than the wages to which they are entitled. The fact that people will often *accept* these wages does not prove that they are not *entitled* to more; it simply shows that they occupy a weak structural position, in which they take the modest rewards they can get because they do not have better options (Mayer 2007). The willingness to accept baseline is problematic, then, inasmuch as it not only may allow privileged actors to extract excessive payments, but also may involve disadvantaged actors settling for much less than they should receive, if their opportunities were genuinely fair. But this, of course, requires an account of what it means for our opportunities to be fair. Our next baseline provides an answer to that question, by connecting payments to the goal of the eradication of poverty.

The anti-poverty baseline

According to 'minimalist' views about global justice (see Armstrong 2012), everyone should have the ability to escape from poverty (Rawls 1999, Miller 2007). In some cases, this means the privileged should provide the disadvantaged with positive assistance aimed at helping them to find their way out of poverty. In other cases, it means they must be vigilant to ensure that institutions, practices, and policies do not make it harder for people to escape poverty nor take advantage of their unfairly limited opportunities. Minimalism as a view about global justice has not been applied explicitly to conservation issues, unlike other important issues such as international trade, migration, and climate justice. But its implications appear to be fairly clear: if people *ought* to have realistic opportunities to escape from poverty, and if conservation measures threaten to cut away their only realistic path for doing so, that would be unjust. Minimalists can therefore support an anti-poverty baseline, which rules out conservation outcomes that cause people to fall into poverty or that diminish their chances of escaping it.

Minimalists can also object to cases of exploitation, which involve some actors taking unfair advantage of others. David Miller (2009), for example, argues that outcomes can be exploitative if they arise in conditions of significantly unequal bargaining power, and if they therefore grant people less than they are entitled to. For the minimalist, problematic outcomes would presumably include those that avoidably left people in conditions of poverty. For example, imagine a situation where the prevalent wages in a region are $1 per day, but these wages leave people in serious poverty. Perhaps the ability to lead a decent life would demand that everyone received at least $2.15 instead (the World Bank's current extreme poverty line), or perhaps it would require even more. The advocate of a status quo baseline will see nothing wrong with a situation where conservation projects granted people incomes of $1, if that is the prevalent wage in a region. But minimalists can argue that it would be exploitative for conservation organizations to leave people with an income of $1, simply because they had the power to do so. While conservation outcomes can sometimes be objectionable because they push people into poverty, in other cases they can be objectionable simply because they avoidably leave people in poverty, where this involves denying them a fair return for their sacrifices. For minimalists, those who determine conservation policies can have a duty to avoid taking unfair advantage of the unequal structural position people find themselves in, and avoidably leaving people below a reasonable poverty line is a good example of such an unfair practice.

The anti-poverty baseline, then, delivers quite distinctive guidance in practice. In some cases, the anti-poverty baseline will actually be less demanding than the status quo and willingness to accept baselines. Imagine a farmer who earns an income far above the global average. If a conservation project requires him to forego that income, the status quo baseline suggests he should be given help in obtaining a similar income in some other way. The willingness to accept baseline is likely to lead to similar results, assuming that the farmer will not be prepared to accept a reduction in income. But the anti-poverty baseline focuses on ensuring that the farmer is not left in poverty, and so it is not clear that it would object to him ending up worse off than average as a result of the conservation project, so long as he stays above the poverty line.

In other cases, I have shown that adopting an anti-poverty baseline will suggest that people are owed *more* than they are used to receiving, and in those cases the view will be more demanding than the status quo or willingness to accept versions. This may be because outcomes that leave people in poverty can count as exploitative, where they result from unequal bargaining power, or because we have a positive duty to help people escape from poverty. This marks a clear difference between the anti-poverty baseline and the two previous baselines considered here. Unlike the status quo and willingness to accept baselines, advocates of the anti-poverty baseline need not accept outcomes where people are left below the poverty line.

The anti-poverty baseline is in this respect an advance, insofar as it avoids the conflict between conservation and poverty: no one will be kept in poverty as a result of the need to engage in conservation, and people's poverty should not provide an opportunity to interact with them on exploitative terms. Global justice scholars have made a somewhat parallel argument in the case of climate change, arguing that if it is possible to meet the costs of mitigation without pushing anyone into poverty—or removing their chances of escaping it—then that is what ought to be done (Moellendorf 2014). The same argument has not been extended to the biodiversity crisis, but it can be. As such, an anti-poverty baseline is an indispensable part of any account of conservation justice. Nevertheless, the anti-poverty approach is not sufficient as a fair baseline, because of the way it treats people above the threshold of poverty. For the minimalist, injustice occurs when costs are unfairly loaded onto the shoulders of people below the poverty baseline. But minimalists are not concerned with the distribution of benefits and burdens above the poverty baseline (Armstrong 2012). In fact, minimalism implies that there would be no injustice even if *all* of the opportunity costs of conservation were loaded on the shoulders of people who live just above the poverty threshold, so long as this does not push them into poverty. Implausibly, it implies this would be a fair outcome even if there were many other people with far greater capacity to absorb burdens. In the case of climate change David Miller (2007) has suggested, to the contrary, that above the threshold of poverty, burdens ought to be shared equally. But making that claim takes him away from the minimalist position and onto the territory of an egalitarian account (Armstrong 2012: 198–9). Strictly speaking, the structure of the minimalist view refuses to accept that facts about people's *comparative* opportunities (above the poverty threshold) are relevant when we come to evaluate a particular distribution. Instead, it focuses solely on ensuring that everyone has the opportunity to escape poverty.

However, theories of justice should also concern themselves with what happens above the poverty baseline. It would be unjust if the most advantaged loaded the costs of conservation onto the shoulders of people who were not poor, but who were much worse off than themselves. Ensuring that conservation does not leave others in poverty is important. To share burdens fairly above the poverty baseline, we need to turn to an egalitarian baseline.

The egalitarian baseline

Prominent egalitarian theories of global justice maintain that, other things being equal, people the world over ought to have roughly equal prospects in life, in some metric to be specified (see e.g. Caney 2005). The alternative is to argue, implausibly, that it is fair for some of us to have worse prospects than others simply because of where we are born, or because of the position of our country in the global economy. If our lives have equal moral value, it is hard to see how such a case could be made successfully. The idea of 'equal prospects' has to then be unpacked in some way, and my suggestion is that it is best understood in terms of people's access to well-being (Armstrong 2017a), where well-being is understood in a hybrid fashion as substantially objective but also as possessing an irreducibly subjective dimension (see Wall and Sobel 2021). What matters morally is our ability to lead healthy, fulfilling, and reasonably autonomous lives wherever we happen to live. Like minimalists, egalitarians therefore care about ending poverty, because poverty seriously jeopardizes our ability to lead flourishing lives. But distinctively, egalitarians are also concerned with people's *comparative* prospects, even above a baseline of poverty. It is wrong if some have worse prospects in life than others, through no fault of their own, even if they do not find themselves in poverty.

According to an egalitarian baseline, opportunity costs should be calculated in terms of shortfalls from an equal sustainable standard of living. This baseline would suggest that an allocation of conservation burdens is unjust if it prevents people from achieving an equal sustainable level of well-being, or if it involves exploitative transactions that take advantage of people's unequal opportunities or access to resources. This represents a more demanding standard than the anti-poverty baseline, which would only judge an allocation to be unjust if it prevented people from achieving a minimal or decent standard of living or exploited their inability to secure such a living. The egalitarian baseline is more compelling, though, because setbacks to our access to well-being are objectionable not only when they leave people below some suitable poverty line, but also when they mean some of us have *better* prospects than others.

This means that conservation projects should not push people into relative disadvantage, nor worsen their chances of escaping from it. But it also means that conservation projects should not exploit people's unjustly diminished prospects. Even if they do not fall below some reasonable poverty line, people in the global South often suffer from relative disadvantages, such as inferior access to productive capital, weak institutions, more modest education opportunities, limited access to healthcare, geographical disadvantages, and so on. Often, these impediments will have been created or intensified by a legacy of colonialism and unjust aggression, and by global economic institutions that continue to offer a raw deal to people in the global South. It might well be that, even in the absence of any impediments to their prospects imposed as a result of unfair conservation policies, many of the worst-off people in the world would have struggled to achieve anything close to an equal sustainable standard of living. But conservation projects can still be unjust when they take unfair advantage of people's inferior access to well-being. Projects that take advantage of people's unfavourable structural positions in order to pay less than they *ought* to get—which egalitarians will define in terms of an equal sustainable standard of living—violate a duty not to exploit those who are worse off through no fault of their own.

The egalitarian should also want to draw linkages between conservation policy and the wider project of ameliorating the disadvantages faced by the world's poor, by using conservation as an opportunity to promote *better* opportunities than they would otherwise have enjoyed. We live in a deeply unjust world and are likely to do so for the foreseeable future. In moving towards a more just world, many tools are likely to be important—including, *inter alia*, trade policy, tax policy, lending and investment, aid, migration policy, and climate policy (Armstrong 2019a, chapter 3). None of these is likely to be sufficient by itself. As such, the egalitarian has reason to seize any additional opportunities that promise to shift resources in the direction of the disadvantaged and should view with favour any permissible policies that would grant them more control over their own destinies. The egalitarian will argue not only that the allocation of conservation burdens *should not worsen* opportunities to achieve an equal sustainable standard of living, but also that conservation policies should *promote more equal* access to such a standard of living. He or she should envision conservation policy as an integral part of a movement to secure a more equal world order.

The egalitarian baseline and the biodiversity crisis

I have shown that opportunity costs are not a neutral category but rather are inevitably calculated in relation to one baseline or another, and that all of these baselines unavoidably involve moral judgements. In real-world conservation projects, where opportunity costs are shared at all, the default assumption appears to be that policy makers should employ either a status quo or willingness to accept baseline (which is not to say that real-world conservation payments always satisfy the status quo or willingness to accept baselines in practice: in the aforementioned Ankeniheny–Zahamena Corridor project in Madagascar, payments appear wholly insufficient to restore people to the *status quo ante*). But I have argued that we should reject those baselines, because they would render conservation policy makers complicit in the severe disadvantage many people wrongly face. Adhering to such baselines could also mean that conservation policies are exploitative if they take unfair advantage of people's unfavourable structural position to give them less than they are entitled to when their lives are disrupted by those policies. I suggest that this is a good example of the fact that we ought to evaluate the impact of conservation policy in light of broader concerns about global justice. Minimally, I have argued that policy makers should commit to an anti-poverty baseline, with the implication that conservation policies should neither make it harder for people to escape from poverty nor exploit their poverty. But the anti-poverty baseline, while important, is not enough. Nor should conservation policy make it more difficult for people to achieve an equal sustainable standard of living; this means that conservation policies that merely avoid locking people into poverty are insufficiently demanding. More ambitiously, I have suggested that conservation policy could aim to shift the opportunity structure that people face, *improving* their prospects and thereby bringing the goal of an equal sustainable standard of living closer to fruition.

I recognize, however, that this is a challenging view, and in this section I address a worry about it. Specifically, it might be thought that making conservation policy a vehicle of global justice is inappropriate, because poverty and biodiversity loss are distinct problems that ought to be dealt with separately (Terborgh 1999). More specifically, it might be argued that linking the two challenges will only slow urgent action to tackle the biodiversity crisis (Kinzig et al. 2011). If urgent action is required now, then making individual conservation projects more costly (as both the anti-poverty and egalitarian baselines would do) will undercut collective responses to the biodiversity crisis. In so doing, it might make everyone, including the poor, worse off.

To reinforce that point, some might invoke a distinction between justice as fair burden sharing, and justice as harm avoidance (Caney 2014). Justice as fair burden sharing asks what an ideal allocation of burdens would be when it comes to tackling some common problem. Justice as harm avoidance, by contrast, begins by suggesting that urgent action is required in order to tackle a problem, and asks how we might bring that about speedily and effectively. In some cases, the two approaches may deliver similar conclusions. But in others, justice as harm avoidance can condone *unfair* allocations of burdens if this is necessary to avoid still greater harms further down the line. In the case of climate change, for instance, a politician might suggest that the only practical way to get very affluent people to stop driving SUVs is to offer them generous incentives to trade them in for electric vehicles, even if the drivers in question can and should bear these transition costs themselves. Although affluent drivers do not 'deserve' such incentives, neither do future people deserve the very great harms that will come about if dangerous climate change is not avoided. *If* drivers will not give up their SUVs without incentives, and *if* moving away from SUVs is necessary in order to avoid climate catastrophe, then the harm avoidance argument might suggest that it is better to give in to their demand for incentives (even if it is an unreasonable one), rather than let climate disaster unfold. In the case of biodiversity loss, the harm avoidance argument might be taken to imply, for example, that policy makers should pay very affluent people to set aside farmland, even if they could easily bear the costs of doing so themselves. That would introduce some unfairness—but doing so might be necessary in order to prevent much greater harms to innocent others. If we focus on justice as fair burden sharing, such policies look unpalatable. But if we focus on justice as harm avoidance, such policies might be justified in the interests of avoiding still greater unfairness (towards people who would otherwise be harmed in the future as the biodiversity crisis unfolds).

If we emphasize harm avoidance, it could also be argued that policy makers are justified in giving some people *less* than they are entitled to if this is the only feasible way to achieve action on a sufficiently large scale, in the right time frame and with the resources available. Requiring poor locals to bear some of the opportunity costs of biodiversity conservation is morally objectionable, and it might well be exploitative; however, allowing environmental catastrophe to unfold, we might say, would be even worse. On grounds of harm avoidance, it might be argued that it would be a mistake to operate an egalitarian baseline for conservation costs, *if* this means that the biodiversity crisis is not averted, or even if it becomes significantly less likely that it will be averted.

How should we assess this argument about harm avoidance? I agree that harm avoidance can sometimes mean that policy makers are justified in imposing policies that do not distribute burdens fairly. But this does not mean (and Caney would not himself claim) that we should place considerations of fair burden sharing to one side entirely. There are four reasons why it is important to be clear about what fair burden sharing would look like, even if urgent action to arrest the biodiversity crisis is absolutely vital.

First, even if the only way we can tackle a pressing problem involves imposing an unfair allocation of burdens, it is important to know what a fair pattern of burden sharing *would be*. For one thing, we should recognize that political or material circumstances can change over time, such that policy makers *can* move closer to justice. To do so, they need to know what a just allocation of burdens would look like. For another thing, even if we cannot move closer to justice, it is important that we recognize when we have departed from its demands. Disadvantaged people might be owed a profound apology if they receive less than they are owed when conservation policies are set. If such an apology was sincere, it would imply a commitment to make policies fairer, and to compensate past losses, if and when it becomes possible to do so.

Second, it might *not* be the case that fair policies are impossible or even politically infeasible. Given the immense resources available in the world, it might well be that applying an egalitarian baseline to conservation policy would be eminently possible. For example, the existing funding channelled towards conservation projects pales in comparison to the funds earmarked for new fossil fuel extraction schemes and other infrastructure projects (Watson et al. 2021). Diverting just a small share of these immense resources might allow policy makers to give affected people the support they are entitled to. Even those funds that are already earmarked for conservation are poorly targeted, with the vast majority of funding spent within the global North, despite the fact that the global South contains most biodiversity (Stark et al. 2021: 2). This requires urgent remedy. The question of political feasibility, of course, is a moveable feast, and the point of arguments about global justice is often to shift consensus about what is morally acceptable (Gilabert 2017). But even if fair burden sharing *is* in some sense politically infeasible, it is important to register that this is often not because of an overall lack of resources, but because of their hugely unequal distribution, and because of the hold that the very well-off have on the exercise of political power. Those who *make* a particular allocation of burdens infeasible might be acting wrongly, and if so, it is important to recognize that fact. Other actors can work to reveal their advantaged position, and their pernicious effect on policy making.

Third, imposing unfair policies will often serve to undermine commitment to conservation projects in the long run. An allocation which is widely recognized as fair, by contrast, can expect to garner greater public support (Martin 2017: 37, Hamann et al. 2018: 70). To the extent that this is true, it cuts against the claim that we can have pragmatic reasons for imposing an unjust allocation of burdens. Imposing an unfair allocation of costs may not in fact be politically sustainable over time, insofar as it undermines support for conservation policies. More broadly, conservation pressures are often generated within a highly unjust global economy, in which the poor are locked out of valuable opportunities and then, to add insult to injury, asked not to exploit local natural resources in the interests of conservation. In such cases it is the assumption that locals should cooperate with conservation projects which might be better labelled 'unrealistic'. Offering them better opportunities to advance their well-being would not only be fair in its own right, but also would make them more likely to cooperate. Giving them an appropriate degree of control over the projects concerned could help still more.

Fourth, it is important not to underestimate the extent to which conservation policy not only *can*, but in practice *does*, internalize goals of global justice. Take, for instance, the transfers made under the Payments for Ecosystem Services (PES) framework. As I noted in Chapter 3, we have reasons for caution about the PES framework as a way of conceptualizing our moral relation to conservation projects. But it is interesting to note nevertheless that PES often *does* function as a kind of redistribution, easing poverty, and hence pressure on local resources (Büscher and Fletcher 2020: 196). According to

Martin (2017: 88), 'most operational PES schemes have far less to do with markets than is typically assumed—and in some cases rather more to do with governments and other agencies seeking positive action to redistribute costs and benefits' (Martin 2017: 88). My claim is that we should be explicit about the justice goals that conservation policy ought to embody, rather than allowing them to emerge in an ad hoc fashion. Justice goals should be explicitly designed into conservation policies from the start, both in terms of fair burden sharing and fair participation. But in order to do that, we need to know what fair burden sharing would look like. Establishing the right baseline for opportunity costs is an important part of that project.

Implications for conservation practitioners

Finally, I will now address the implications of my argument for conservation practitioners—people who design, fund, and implement conservation projects. It is helpful here to emphasize the philosophical distinction between negative and positive duties. Negative duties include duties not to harm other people nor to treat them wrongfully. A small-scale example would be the duty not to push someone who cannot swim into a swimming pool, or to make it harder for her to get out if someone else has already pushed her in. In debates about global justice, it is commonly agreed that negative duties are held by everyone, regardless of their location or social position (Miller 2007). It is wrong for me to push someone who cannot swim into a swimming pool, and it is similarly wrong for anyone else to do so, whoever or wherever they happen to be. Since negative duties are held regardless of distance or the particular roles we inhabit, it is plausible that they apply to conservation practitioners, too.

One of the most important negative duties, which has been the subject of much discussion in the literature on global justice, is a duty to avoid pushing people into poverty, and, relatedly, a duty to avoid making it harder for people already in poverty to escape from it (see e.g. Pogge 2002, Miller 2007). Another is the duty not to exploit people by taking unfair advantage of their limited opportunities. These duties surely apply to anyone involved in conservation policy, funding, research, or implementation, all of whom should be vigilant that, if at all possible, their decisions do not make the problem of global poverty worse nor exploit people's straitened circumstances.

Harm is not limited to pushing people into poverty, exploiting their poverty, or making it harder for them to escape from it, however. Conservation policies are objectionable not only when they lock people into (or take advantage of) poverty or absolute disadvantage, but also when they lock people into (or take advantage of) comparative disadvantage. This is the clear implication of the egalitarian baseline, which condemns conservation policies that make it more difficult for people in poor countries to catch up with the rest of the world. Observing their negative duties means that conservation practitioners must also take care that their interventions do not make it more difficult for the poor to attain an equal sustainable standard of living nor exploit their unjustly limited opportunities.

Positive duties, by contrast, typically involve offering active assistance to others in need, as opposed to merely not harming them (Singer 1972). For instance, on seeing that someone has been pushed into a swimming pool, a bystander might have a duty to help rescue them. In debates on global justice, positive duties are often connected with the capacities of particular actors to make a difference. For instance, it has been argued that those who *can* help to reduce global poverty ought to do so, at least when doing so involves them taking on no more than moderate costs (Barry and Øverland 2016). I argue, further, that those who can assist in the transition to a more equal world should join that effort, even if it involves them taking on some costs. The question then becomes who actually possesses the capacity to make a difference, and what constitutes a fair division of labour. I do not claim here that conservation practitioners or those who determine conservation policies and priorities are the *only* actors with the capacity to make a difference. There will be many actors who can help improve opportunities for people caught in the conservation/development dilemma, including national governments in the global South, governments in the global North, aid agencies and non-governmental organizations, and international institutions such as the World Bank, International Monetary Fund, and World Trade Organization. Among all of these actors, conservation practitioners and policy makers are surely not the most powerful. What we should expect of such actors will depend on the time and other resources they possess, and the likely costs to them of pressing for change (Caney 2012a). Much may depend on their class, race, and gender: we should avoid the double bind of asking actors who are already disadvantaged to bear special burdens in reducing injustices, and should seek instead to place greater onus on members of more privileged groups (Yankah 2019). In many cases, it may be that the part any particular actor can and should play in discharging our positive duties towards the world's poor will be quite small. Nevertheless, there is a whole series of ways in which academics and other professionals can contribute to tackling problems like global poverty (Caney 2012a), and it is likely that they can make *some* difference. For example, they might be able to contribute by arguing for a more robustly anti-poverty and even egalitarian approach to conservation policy, which avoids leaving those affected by conservation interventions locked into poverty, and even assists them in accessing a fairer and more equal standard of living. They might contribute to debates that, over time, could shift the policies of conservation organizations, governments in the global North, and even international institutions. Even if their capacities to change the world are probably quite limited, conservation practitioners can still be important allies in the struggle for global justice.

Conclusion

Simon Caney (2012b) argues powerfully that discussions of our response to the climate crisis should not play out in isolation from issues of global justice, because if they do, policy makers risk entrenching existing injustices. To the contrary, debates about climate change ought to be integrated with broader discussions of global justice in order to ensure that our responses to the climate crisis are fair. I believe that the same holds for debates on biodiversity conservation. In this chapter I attempt to illustrate the importance of that kind of integration by arguing that the baseline for calculating opportunity costs can be defined in a number of different ways, and that we have moral reasons for preferring some baselines rather than others. The danger of applying a status quo or willingness to accept baseline to conservation funding schemes is that this may either entrench or exploit distributive injustice, given that people's actual incomes, and expectations, are shaped within a context of profound injustice. At a minimum, justice requires that policy makers apply an anti-poverty baseline, which would avoid locking those affected by conservation into poverty. More ambitiously, I argue for an egalitarian baseline, connected to an equal sustainable standard of living. I also engage with a powerful objection to the egalitarian baseline, which suggests that putting it into practice might slow progress in tackling the biodiversity crisis, in effect placing much greater demands on a limited conservation budget. It may or may not be true that operating a more demanding baseline will make progress in tackling the biodiversity crisis less likely. But even if it was true, it would be important to recognize that, in their drive to tackle the biodiversity crisis, policy makers were imposing an *unfair* distribution of conservation burdens. Over time, they ought to work to make it possible to apply a fairer—more egalitarian—baseline. In doing so, conservation policy makers can be important allies in the struggle for global justice.

5
Justice and biodiversity offsetting

Biodiversity offsetting is a specific, and increasingly influential, form of trading. Paradigmatically, the practice involves a 'developer' (A),[1] paying another actor (B) to restore biodiversity on a site B controls (a 'restoration offset'), or to preserve biodiversity that B would otherwise have destroyed (an 'avoidance offset'). A then gains the permission to destroy an 'equivalent' amount of biodiversity on a site they control themselves: in the prevalent terminology, any destruction at the 'impact site' is counterbalanced by conservation at the 'offset site'. The practice has its origins in a US wetland regulation of the 1970s. Initially, developers were required to avoid or at least minimize damage to wetlands; lobbying from commercial actors helped persuade policy makers to render regulatory pressures more flexible (Damiens et al. 2021: 174). The introduction of offsetting meant developers were required to ensure that their activities did not diminish the extent of wetland *overall*, but were permitted, if they desired, to destroy wetland they owned so long as they paid others to conserve wetland elsewhere, where the economic returns to land might be lower (Martin 2017: 137).

This US policy was an example of a 'compliance' or 'regulatory offset', where the purchase of an offset is *required* in order for developers to gain official approval for changes in land use. Compliance offsetting schemes have since come to be a very familiar domestic policy tool within the UK, Australia, and many EU member states, where they are now strongly associated with 'no net loss' biodiversity policies.[2] They have also been adopted in many countries in the global South, including Brazil, India, and China (Bidaud 2017). On one estimate, offset sites now cover land equivalent to the total area of Bangladesh (Bull and Strange 2018), with the market for compliance offsets worth an estimated US$10 billion.

---

[1] I use the term 'developer' in this chapter simply to mean an actor who would like to destroy biodiversity in order to advance some economic goal.
[2] The term 'no net loss' emerged in the late 1980s. It is important to recognize that the close association between biodiversity offsetting and no net loss is nevertheless contingent. In principle, policy makers could adopt no net loss policies but forbid biodiversity offsetting, or allow offsetting but set a baseline for biodiversity preservation that is more or less demanding than no net loss (McLaren and Carver 2023).

Global Justice and the Biodiversity Crisis. Chris Armstrong, Oxford University Press.
© Chris Armstrong (2024). DOI: 10.1093/9780191888090.003.0006

These compulsory schemes have been accompanied by versions that are more voluntary in character. Voluntary schemes involve individuals or corporations *choosing* to buy offsets, with that purchase enabling them to make claims about the 'biodiversity-neutral' or even 'biodiversity-positive' nature of their economic activities. For example, Colombian developers can now choose to buy Voluntary Biodiversity Credits (VBC) from the El Globo Cloud Forest Habitat Bank in the Andes, which protects habitat that is home to, *inter alia*, the endangered Spectacled Bear. Each VBC covers ten square metres of habitat, and costs US30$.[3] If they buy sufficient VBCs, Colombian companies might be able to claim that their destructive activities elsewhere in the country have been offset by habitat protection at El Globo. Viewed from one perspective, companies that buy voluntary offsets are 'pre-emptively managing business risks' including potential reputational damage (Koh et al. 2019: 679). Viewed from another, they might be responding to a lack of effective compulsory policies by ensuring that their own actions, at least, do not cause a net loss of biodiversity.

The rapid expansion of offsetting schemes over recent years is remarkable. One reason why it is remarkable lies with the fact that conservation practitioners and policy makers are often explicitly committed to a 'mitigation hierarchy'. This decision procedure suggests that, when biodiversity is threatened, the avoidance of destruction should be the first resort. If destruction cannot be avoided, actors should aim at its minimization, and to the extent that this also fails, they should then aim at its remediation *in situ*. Only once *all* of those options have been exhausted, it is widely believed, should they turn to offsetting (Arlidge et al. 2018: 337). However, the dramatic growth in biodiversity offsetting schemes suggests that offsetting has actually become a first, rather than last, resort for many developers—a process that, in the years since the US wetland scheme was introduced, policy makers in many countries have accepted and indeed facilitated. There is now considerable evidence that offsets are displacing avoidance and minimization as responses to biodiversity loss (Ives and Bekessy 2015).

This chapter assesses the normative implications of this trend. It begins by briefly showing why biodiversity offsetting matters from the point of view of global justice. It then moves on to consider the positive case for offsetting. Although the focus here is on biodiversity, at several points I also draw comparisons with the practice of carbon offsetting, which has received

---

[3] https://en.terrasos.co/nota-cbv-climatrade-y-terraso (Accessed 11 November 2022).

considerably more attention within political theory to date.[4] However, I argue that biodiversity offsetting is in several ways a more problematic practice than carbon offsetting. Specifically, we have grounds for concern about offsetting both in virtue of its impact on human societies, and in virtue of its impact on non-human animals. These reservations apply at both ideal and non-ideal levels: while some of these concerns could in principle be alleviated by way of careful policy design, others appear likely to apply in more or less any situation where offsetting is adopted. I conclude by sounding several cautionary notes about the role that offsetting has begun to play in biodiversity governance, which, taken together, suggest that to date reliance on offsetting may well have done more harm than good.

Biodiversity offsetting and global justice

Before we proceed further, I will highlight why biodiversity offsetting might be seen as interesting—and important—from the point of view of global justice. This is not immediately obvious because to date, markets in biodiversity offsets have remained 'almost entirely local' in character (WEF 2022: 4): that is, they involve damage to biodiversity within one country being 'compensated' by restoration or avoided loss within that same country. Indeed, there is a fairly widespread norm that biodiversity offsets should involve the destruction of biodiversity being counterbalanced by the conservation of biodiversity that is ecologically similar, and this is widely thought to speak in favour of within-country offsetting.

Nevertheless, there are (at least) five reasons for connecting biodiversity offsetting to debates on global justice and, taken together, they explain why an analysis of offsetting clearly merits a place in this book. First, the proliferation of offsetting schemes at the national level is partly a consequence of its *promotion by global institutions*. The World Bank, for example, now requires any major infrastructure projects that it funds to be consistent with the goal of no net loss of biodiversity, and this alone has done much to boost the market in offsets (zu Ermgassen et al. 2019b: 308). Text appearing to endorse biodiversity offsets has also been included in recent decisions issued under the Convention on Biological Diversity framework. Even if offsetting schemes are national in scope, therefore, the proliferation of those schemes is a phenomenon that global actors have encouraged and facilitated,

---

[4] This relative neglect of biodiversity offsetting is somewhat curious, since biodiversity offsetting actually emerged first, and to a large extent provided the conceptual apparatus and business model for carbon offsetting (McLaren and Carver 2023).

and one important question is whether they should continue to do so. Second, many of the *actors* that participate in offset markets are transnational in character. For instance, Rio Tinto (a British–Australian mining corporation with headquarters in London) famously offset the biodiversity impacts of its ore-mining activities in Madagascar by protecting habitats elsewhere in that country.[5] Major global companies like Nestlé and Bayer, meanwhile, have recently made claims about the 'nature-positive' character of their activities, which depend in large part on the existence of biodiversity offsets around the world (Ferns 2022). I show, however, that the activities of such transnational actors can worsen problems of exclusion and dispossession within the communities in which they operate. Third, the presence of offsetting has *consequences for the global economy*. For instance, the availability of offsetting can be expected to lower the price of 'development' in the countries where it is practised. Among other things, this might cause countries that insist on the avoidance or minimization of biodiversity loss to lose out on investment, leading to downward pressure on environmental regulation globally. Fourth, I show that offsetting can have implications for the *distribution of opportunities between* communities. Fifth, it may be that genuinely transnational offsetting schemes *will emerge* in the coming years. For example, the European Commission has spent considerable time developing a biodiversity framework that commentators suggest will soon allow biodiversity destruction in one member state to be offset by conservation in another.[6] Some have argued that biodiversity markets *should* come to resemble carbon offsets more closely, not least by becoming truly global in scale.[7] In fact the world has already witnessed the emergence of 'biodiversity credit' schemes that are transnational in character. While their supporters claim that these schemes do *not* involve offsets, it is also clear that, just like offset schemes, they involve corporations funding conservation abroad and then making claims about the 'nature-neutral' or even 'nature-positive' credentials of their business operations as a whole (WEF 2022: 6–7). For all of these reasons, it is important to assess the global justice implications of the proliferation of offsetting.

---

[5] Although Rio Tinto eventually abandoned its company-wide commitment to 'net positive impact on biodiversity' (Global Forest Coalition 2022: 2).
[6] https://greenfinanceobservatory.org/wp-content/uploads/2022/07/EU-nature-restoration-targets-brief-2.1-final.pdf (Accessed 17 November 2022).
[7] https://blog.toucan.earth/biodiversity-the-next-frontier-for-tokenized-markets/   (Accessed   16 November 2022).

The case for offsetting

As noted, the practice of biodiversity offsetting has been widely institutionalized. The same, of course, can be said for carbon offsetting. The United Nations' Clean Development Mechanism, for instance, allows countries to obtain extra emissions entitlements by funding climate mitigation projects abroad: their extra emissions are 'offset' or excused by matching emissions reductions generated by those projects. Just as biodiversity offsetting has come to be closely associated with the goal of no net biodiversity loss, carbon offsetting has become closely associated with 'Net Zero' climate targets. Individual corporations are now facing considerable pressure to commit to the goal of Net Zero emissions by 2050. Many of them—especially, but not only, in industries like aviation—have made Net Zero commitments that envisage a large portion of their future emissions being offset (Armstrong and McLaren 2022).

Why, though, might it be a good idea for policy makers to allow—or even to encourage—offsetting? Two general reasons tend to be provided within the literature on carbon offsetting, and appear to be readily applicable to the biodiversity case as well. First, consider efficiency. Allocating resources efficiently is morally desirable, because it means that actors can advance their various ends more completely than they could otherwise with the same quantity of resources (Satz 2010: 22). Governments, however, typically possess distinctly limited information about the various purposes to which particular resources could be put. Among other things, they may not know where it is cheapest to make emissions reductions (Caney and Hepburn 2011: 205, Gosseries 2015: 95) or to avoid biodiversity loss. For governments to impose uniform emissions reductions on actors, for instance, would therefore be comparatively wasteful. A policy of offsetting—where one private actor can pay another to engage in emissions cuts on his or her behalf—should be more efficient, and therefore reduce the overall social costs of meeting any macro-level emissions reduction target. In principle, this should mean that greater emissions reductions are feasible within a given time period— *if* policy makers choose to ratchet up their ambition. The same should be true for biodiversity offsetting. Consider, for instance, a policy that stipulated that each actor must preserve precisely 80 per cent of the biodiversity he or she controls. Compared to a policy along those lines, a policy that embraced offsetting would be more efficient. Pursuing biodiversity conservation where it is cheapest to do so means conservation can be obtained with fewer resources, leaving more scope to pursue other valuable goals, and potentially making it possible to establish more ambitious macro-level conservation targets.

Second, consider the value of freedom. The practice of offsetting opens up more options for actors, when compared to a system in which those actors are given binding and unwavering targets to reduce either emissions or biodiversity loss. Under a system of carbon offsetting, for instance, some actors may well choose to cut their own emissions so as to meet precisely whatever quota is handed to them. Meanwhile, some may prefer to pursue even more aggressive mitigation in order to sell carbon offsets. Others can then continue to emit beyond their quota if they wish to do so, so long as they are prepared to pay the going price of an offset. It is this freedom, of course, that allows efficient solutions to common problems to be found, compared to command-and-control policies. But freedom is also valuable in its own right: there is simply 'value in allowing individuals to make their own choices' (Caney and Hepburn 2011: 206). Given that what matters is policy makers' ability to secure *overall* emissions reductions, the value of freedom speaks in favour of embracing, rather than forbidding, offsetting, where this is compatible with any overall climate goal. Again, precisely the same argument could be made about biodiversity. Biodiversity offsetting also enhances freedom by allowing people to choose, if they prefer, to specialize in conserving biodiversity, or to engage in more destructive activities, consistent with whatever overall conservation target has been adopted. This is of course a comparative judgement. Chapter 6 looks at some of the negative social impacts that regularly arise when protected areas are imposed on local people without their consent. An offsetting approach in which participation on the offset side was genuinely voluntary—in which people could freely *choose* to offer up local areas as offset sites—would better protect freedom, when contrasted with a command-and-control approach along those lines. Developers would be freer, too, if they then had the option to transform impact sites, provided they paid the requisite price.

The positive case that is typically advanced is clear: given any macro-level target for either emissions reductions or biodiversity protection, allowing offsetting to flourish as a practice will extend the space of freedom, giving economic actors more options than they would otherwise possess; partly for that reason, it will allow any given target to be met more efficiently. Allowing some actors to specialize in biodiversity-destroying activities and others to specialize in biodiversity restoration, for instance, should reduce both the average and overall price of conserving biodiversity. In a world where conservation is forced to compete with other priorities, this could be considered a good thing.

Impact on biodiversity

Despite these ostensible advantages, the practice of biodiversity offsetting also prompts serious normative worries. Some of these are familiar from discussions about carbon offsetting, but some of them are relatively distinctive, and some of those that do apply to carbon offsetting apply with even more force in the biodiversity case. In this section I sketch three grounds for concern about the impact of offsetting on biodiversity, or the creatures that constitute it, before considering worries about global justice and injustice in the next section.

Normalizing biodiversity loss?

The success or failure of biodiversity offsetting schemes is judged, by policy makers, in relation to an overall projection of how much biodiversity loss is expected to take place over time. This is the 'baseline' for biodiversity loss. Whichever baseline is selected, adding offsets to the picture means that actors can choose to trade opportunities, just as they can under a carbon offsetting scheme. Success, for a particular scheme, means that the amount of biodiversity loss is no more than would have been expected, according to the baseline selected, if impact sites had not been developed at all.

The selection of a baseline is, therefore, an absolutely crucial political decision. In practice, regrettably, policy makers often leave the details of their chosen baselines rather implicit when establishing offsetting schemes, leaving outsiders with the difficult task of extracting them from the minutiae of policy documents or technical appendices (Maron et al. 2015b). However, various options are possible. One is to adopt what we might call a *static* baseline. Here, any net decline in biodiversity over time will be counted as a loss, and any net increase over time will be counted as a gain. Another option is to choose a *positive* baseline, which assumes a certain rate of increase in biodiversity over a given period and calculates gains or losses in relation to that. A *negative* baseline, meanwhile, assumes a certain rate of biodiversity loss over time, and permits offsetting in relation to that projection.

By way of example, a variety of Australian offsetting schemes have assumed baselines of continued biodiversity loss of up to 4 per cent per year

(Maron et al. 2015b: 504). Actors then engage in offsetting *relative to* this (negative) baseline. An actor who destroys 4 per cent of the biodiversity they control need not engage in offsetting, in this case. An actor who destroys 6 per cent of the biodiversity they control can purchase an avoidance offset, for example, from an actor who has 'only' destroyed 2 per cent. Many no net loss schemes operate a negative baseline of some kind, assuming as their 'reference scenario' that biodiversity will continue to decline at a specified rate.

The frequent adoption of negative baselines has the very important—but somewhat underappreciated—implication that success in achieving (what are widely designated as) 'biodiversity no net loss' or even 'net gain' policies is in fact compatible with continual declines in biodiversity over time (Maron et al. 2018). An assumption of 4 per cent biodiversity loss per year, for instance, is really very steep: it would see more than half of the biodiversity in a locale lost within seventeen years. But the loss of this much biodiversity, and no more, would allow it to be claimed that no net loss had occurred. The adoption of negative baselines—and the widespread claim that they deliver on the goal of no net loss of biodiversity—can therefore normalize high levels of loss (Maron et al. 2015a, Gordon et al. 2015), deflecting attention from the destructive effects of development. This problem could of course be avoided through the selection of a more demanding baseline. Nevertheless, there is widespread concern that developers have 'gamed' offset policies, pressuring policy makers to stipulate unduly pessimistic baselines for biodiversity loss, with the implication that more moderate levels of destruction can count as 'gains' capable of being traded against heavier destruction elsewhere (Ives and Bekessy 2015). One important implication of all of this is that the 'exposure of assumptions about crediting baselines to scrutiny is crucial to avoid unwittingly offsetting biodiversity to extinction' (Maron et al. 2015a: 511). Given that the use of negative baselines in biodiversity offsetting is not widely understood, it may well be that the public in many countries has been led to possess a false degree of confidence that the problem of biodiversity loss is 'under control', and that the adoption of no net loss and even net gain policies by corporations or governments is actually serving to avoid the net destruction of biodiversity when it is not.

These worries apply especially strongly to 'avoidance' offsets, under which actors generate offsets to sell when they step back from, or downscale, plans to destroy biodiversity. In many cases assumed rates of background declines in the future, under given offsetting schemes, are actually pitched higher than the current rate of loss in an ecosystem (zu Ermgassen et al. 2019a: 9). This immediately generates assets for some actors to trade on offset markets, even if no overall reduction in the rate of biodiversity loss occurs. For instance, an actor who intends to destroy 2 per cent of their biodiversity might find that policy makers actually select a baseline of 4 per cent destruction over the same period. If so, they could carry on with their plans, and sell the additional 2 per cent worth of offsets for cash. This generates a degree of moral hazard, because developers will have a clear incentive to pressure policy makers to adopt more pessimistic baselines for biodiversity loss: the more pessimistic the baseline, the more avoidance offsets there are to sell, and the lower the financial cost (to developers) of destroying biodiversity. Meanwhile, a separate but serious problem with the acceptance of avoidance offsets is that it may be difficult in practice to determine when a loss has in fact been avoided: in some instances, sites that do not appear to be under the threat of development have quickly been designated as 'protected' precisely in order to then operate as avoidance offsets (Thorn et al. 2018). This reduces the costs of destruction still further.

It might be said that these worries about offsetting could be assuaged by simply setting more demanding benchmarks for biodiversity loss. Perhaps the problem is not the embrace of offsetting per se, so much as the adoption of negative baselines. Pick a more demanding (static or positive) baseline, and policy makers would no longer risk offsetting biodiversity to destruction. There is some truth to this, but it is also important to look at the impact of offsets holistically by assessing the *overall* effects of the embrace of offsetting (as distinct from the embrace of one baseline or another). One key worry involves what the embrace of offsetting allows governments *not* to do. That is, by embracing offsetting (and claiming that biodiversity destruction has been 'solved' in the marketplace) governments may no longer need to actively focus on or adopt other conservation policies. Although some governments have explicitly stated that the adoption of offsetting policies will not lessen existing environmental protections (see e.g. DEFRA 2019), there is widespread concern that, whereas in theory offsetting ought to be the last resort on the mitigation hierarchy, in practice it has often crowded out the avoidance and minimization of biodiversity loss. On one estimate, less than a quarter of countries now require 'compensation policies' (including offsetting) only once other options in the hierarchy have been exhausted (zu Ermgassen et al. 2019b: 307).[8] Another study has found that 77 per cent of development projects have not properly applied the mitigation hierarchy

---

[8] The situation with regards to carbon offsetting appears to be similar. As Hyams and Fawcett (2013: 93) note, government advice is often that offsetting should be the last resort within the 'carbon management hierarchy'. But there is abundant evidence that many individuals and corporations buy offsets for avoidable activities (ibid).

(Global Forest Coalition 2022: 3). In jurisdictions where offsets are available, it is widely feared that erstwhile environmental protections are indeed more weakly applied (Phalan et al. 2018).

More specifically, it might be that a turn to offsetting makes a reduction in government conservation funding politically palatable, on the assumption that conservation has been effectively delegated to the private sector (Gordon et al. 2015, zu Ermgassen et al. 2019b: 311, Neuteleers 2022: 131). If the embrace of offsetting is accompanied by a reduction in states' commitments to other kinds of conservation projects, this would be regrettable, not least if some of those projects are preferable, in some senses, to offsetting. For instance, public commitments to national parks or reserves often possess a high degree of political 'lock-in', with high reputational costs for any government that attempted to renege on them. Corporate commitments to offset sites, by contrast, might be much less secure over time, given the proliferation of different corporate vehicles, which are individually likely to attract much less public scrutiny if they renege on their commitments (Moreno-Mateos 2015: 557), and the absence of the kinds of chains of accountability that (to a degree) bind elected governments. If the widespread adoption of offsetting allows governments to claim that they have a mechanism in hand to deal with biodiversity loss, this could sow false confidence that no net loss policies are securely 'managing' the problem of biodiversity loss, and dissipate popular pressure for more radical conservation policies (Walker et al. 2009: 149).

Complexity, equivalence, and loss

Because of atmospheric mixing, a given quantity of carbon dioxide emitted in one place will typically come to have much the same impact on climate change as an equivalent quantity emitted elsewhere. When policy makers shift attention to greenhouse gases more broadly, they have to reckon with the fact that different gases make different contributions to climate change over time. Nevertheless, the overall contributions these gases make to climate change are quite readily compared, allowing scientists to adopt measures such as 'carbon dioxide equivalence' over a designated time period. The fact that there is no great scientific problem of comparability in the case of greenhouse gas emissions makes carbon offsetting schemes appear relatively straightforward, in at least this respect.

This is very far from being true in the biodiversity case, however. In Chapter 2, I noted that biodiversity is a multi-dimensional concept, which incorporates variation at the level of genes, species, and ecosystems.

For policy makers to judge that biodiversity losses have been successfully off-set, there must be some metric for aggregating and comparing losses across these different dimensions. However, there is no natural answer to the question of how much species-level variation should be considered equivalent to how much genetic or ecosystem-level variation. To the contrary, this is an inherently ethical question to which many different answers could be supplied. In practice, policy makers put in charge of offsetting schemes have adopted highly reductionist, simplified metrics for comparing units of biodiversity (Moreno-Mateos et al. 2015: 554). In some cases, they have simply used a single dimension of biodiversity, such as the number of species present on a site, as a stand-in for biodiversity more broadly (Ives and Bekessy 2015: 570). In other cases, sites have been compared according to a few key landscape features, with data contributing to a 'weighted habitat score' for impact and offset sites (Marshall et al. 2020: 2). Partly for reasons of data availability, and partly to prevent the relevant calculations becoming unwieldy, these habitat scores tend to incorporate some features (including species richness, species abundance,[9] and the presence or absence of a few selected habitat features) while disregarding others (most notably genetic diversity).

Within the parameters of this simplified approach, many offsetting schemes do at least demand that offsetting takes place within a single habitat type, with deciduous woodland being offset by deciduous woodland, for example. But the same is not true of all offsetting schemes. In England, for example, the official biodiversity valuation metric dictates that some habitats can be matched across types (with, for instance, the protection of bracken substituting for the destruction of grassland) (Koh et al. 2019: 685). Rio Tinto's famous biodiversity offset in Madagascar involved the avoided loss of some types of forest, which compensated for the destruction of other types of forest, a kind of 'flexibility' which is also present in South Africa's offsetting scheme (ibid: 689).

It seems likely that *some* degree of flexibility will be required if offsetting schemes are to operate at all. If policy makers required matched sites to be perfectly equivalent in habitat type *and* genetic diversity *and* species count, say, that might mean few if any qualifying offset sites could actually be found. In practice, offsetting policies appear destined to disregard some ecological specificities, and to treat others as in some sense fungible. Rather than holding offsetting to an unrealistic standard, I simply note that the adoption of

---

[9] 'Species richness' refers to the number of species divided by area, whereas 'abundance' refers to the number of individuals in each species, divided by area. This exemplifies a phenomenon I noted in Chapter 2, whereby measures of biodiversity often turn out to incorporate a concern for the idea of abundance, which is better considered distinct from biodiversity.

reductionist metrics has important normative ramifications. First, because metrics such as weighted habitat scores are simplifying, policy makers *cannot* in fact be sure that there is no net loss of biodiversity when an impact site is destroyed, and an offset site protected. Since the biodiversity metrics employed in offsetting schemes simply do not attempt to measure all forms of loss, claims about the biodiversity-neutral credentials of any given development project must therefore be taken with a healthy pinch of salt.

Second, even if policy makers were to resort to increasingly fine-grained metrics for biodiversity, more or less any concrete case of offsetting is likely to be accompanied by loss on *some* dimensions. The metrics used in biodiversity offsetting schemes do not typically involve a checklist, whereby each dimension of biodiversity in an impact site must be matched on that dimension by the biodiversity found at an offset site. Instead, schemes typically involve weighted average scores, where losses on *one* dimension can be outweighed by gains in *other* dimensions. Over time, economic actors have lobbied policy makers to enhance the flexibility associated with offsets, often in response to a perceived shortage of offset sites in biodiversity markets (zu Ermgassen et al. 2020). In Australia, for example, pressure from industry has led to the adoption of more flexible rules over time, even where this has meant that the loss of members of endangered species can be offset (Ives and Bekessy 2015: 570)—for instance, by the use of a habitat multiplier, whereby more land of a lower ecological 'quality' compensates for the destruction of less land of a higher quality. The embrace of flexibility tends to expand the supply of potential offset sites, and hence to cut the price of offsets to developers. As a consequence, however, governments' stated ecological objectives (to avoid exacerbating risks to protected species, say, or to preserve certain types of habitat) may become compromised in the minutiae of policy design (Calvet et al. 2015).[10]

While there are no doubt examples of better and worse policy, it seems that these issues of commensurability would be extremely difficult to design out entirely, and hence that they would likely apply to *any* offsetting scheme. My point here is simply to clarify the kinds of losses that are likely to be involved in offsetting schemes. Even 'successful' offsetting policies can lead to reductions in certain types of habitats, in the size of certain plant or animal populations, and in genetic diversity, as well as hastening the extinction of some species. In order to embrace offsetting, policy makers must judge

[10] Walker et al. (2009: 149) predicted that economic actors would successfully persuade officials to relax safeguards governing offset schemes over time, claiming that offsetting should therefore be considered 'more vulnerable than pure administrative mechanisms to institutional dynamics that undermine environmental protection'. This prediction appears to have been borne out over the last decade or so.

that those losses are outweighed by gains elsewhere. That in turn requires them to take a highly instrumental attitude towards biodiversity, where quantity or ecosystem function, say, are valued, but where the conservation of individual habitat types, species, and indeed individual animals are not prioritized. Among other things, this approach may be incompatible with showing proper respect for the rights of the individual animals that partly compose biodiversity.

Offsetting harm?

One issue absolutely key to ongoing debates about emissions offsetting is whether we should consider carbon emissions to be harmful at all, provided those emissions are offset. Hyams and Fawcett (2013: 95) argue that we should, and that this is a major problem. After all, we do not standardly think that we can hurt someone, go on to pay a third party to stop hurting someone else, and conclude that our actions are permissible. In response, Placani and Broadhead have argued that we should resist appraising emissions separately, and instead consider emissions and offsets together as a set of actions. This is because our emissions are not individually harmful, but rather make a *cumulative* contribution to warming over periods of years (Placani and Broadhead 2022: 7). Offsetting, though, cancels any cumulative impact before it can occur.[11]

John Broome (2012) has also suggested that the two-part action of [emitting a certain amount + paying someone else not to emit an equivalent amount] does not cause any increase in atmospheric carbon concentrations, and hence is not harmful.[12] On such a view, [emitting + offsetting] does not look morally worse than not emitting in the first place (see also Gosseries 2015: 99). Notably, this defence has been applied even to 'luxury' emissions.[13]

---

[11] For a more fine-grained discussion, see Barry and Cullity 2022, who argue that our appraisal should vary according to the kinds of acts that are combined. Combining emitting with sequestering (for instance, by planting trees) may be morally neutral overall, they suggest. But combining emitting with 'forestalling'—for instance, paying someone else to reduce their own emissions—can still, they suggest, involve wrongly imposing a risk on others. Their distinction between sequestering and forestalling offsets appears to run somewhat parallel to the distinction between restoration and avoidance offsets in the biodiversity case.

[12] More recent work suggests Broome might now phrase that claim differently. Because of the instability of the climate, it is possible that ostensibly similar actions, taken at different times or places, might in fact have quite different effects on people's interests further down the line. Nevertheless, we can still say that our actions cause the same degree of *expected* harm (Broome 2019). A reworked justification of offsetting could state that emitting plus offsetting does not increase expected harm (rather than stating that it does not increase harm *tout court*).

[13] Defined as emissions that are not crucial to anyone's basic rights. For an influential early account of subsistence versus luxury emissions, see Agarwal and Narain 1991.

Though luxury emissions *can* certainly cause harm, they will not do so, the argument goes, if one buys sufficient emissions entitlements from other actors who then desist from emitting (or deploy negative emissions techniques of one kind or another). This argument could, for instance, exculpate the vacation flights of the rich, which simply make no difference to climate change if they are successfully offset (Placani and Broadhead 2022).

Crucially, this *harm-cancelling argument* does not claim that harms are *compensated* when we offset; rather, it claims that harms do not *arise* in the first place when we successfully offset. I make two responses to it. First, even if this argument succeeds in defending the emission of carbon *qua* greenhouse gas, that does not mean that it succeeds in defending carbon-emitting activities considered more holistically. This is because their potential contribution to climate change is only one negative effect associated with those activities. Luxury flights, for instance, do not only contribute to climate change: they also contribute to particulate pollution (Placani and Broadhead 2022: 12), which has enormously negative consequences for both human and non-human life. The same can be said of driving petrol and diesel vehicles, using fossil fuels for domestic heating, and so on. Particulate pollution from fossil fuel use has been shown to have dramatic consequences for human health and well-being (Kim et al. 2015) and also seriously harms other animals. Fossil fuel use also supports extractive industries that have often been associated with violence, repression, and political exclusion (Wenar 2015). It is difficult to even conceive what it might mean to cancel these harms before they occur.

Second, and more pertinently for the argument here, the claim that acting-and-offsetting should be appraised as morally neutral even as a set is significantly harder to sustain in the biodiversity case. Consider again the distinctiveness of emissions. As we know, carbon emissions from myriad points on the earth's surface are quite rapidly mixed together in the global atmosphere. The radiative forcing of the atmosphere might have different *effects* in different parts of the planet, but the *source* of emissions does not appear to be very consequential. Moving an emissions-causing activity from point A to point B on the globe is said to make little or no difference when it comes to overall climatic outcomes; this is sometimes thought to make the argument for carbon offsetting easier (Gosseries 2015: 95). To be clear, it is not the case that emitting does localized harm in place A, which sequestration in place B then counterbalances. Rather, the crucial claim—if arguments for carbon offsetting are to be successful—is that no harm occurs in the first place.

The harm-cancelling argument, though, cannot be sustained in the case of biodiversity offsetting. Offsetting, recall, does not comprise the relocation of biodiversity; if biodiversity were successfully relocated, then there would be no need for offsetting. Offsetting comprises the *destruction* (or at least displacement) of biodiversity in one place, and the emergence (or sparing) of an 'equivalent' amount of biodiversity in another place. This is a significant difference, which has important normative ramifications. The most important is that [destroying biodiversity + offsetting] simply cannot be thought of as a harmless practice, even if [emitting + offsetting] can. This is because destroying biodiversity at the impact site is very likely to involve harming (and likely killing) many individual animals.[14] Even when they are not killed, mammals, birds, reptiles, amphibians, fish, and invertebrates are highly likely to be exposed to stress, hunger, separation from their families, increased vulnerability to predation, and so on.

As such, biodiversity offsetting fails a key test that defenders of carbon offsetting often endorse. Defenders of carbon offsetting typically accept that we cannot, say, kill one human being and vindicate that action by paying someone to save another. This is because killing a human is intrinsically wrong (Broadhead and Placani 2021: 410–11). Their claim is that emissions are *not* harmful in the first place—so long as the appropriate offset has been bought from another actor, who then does not emit herself. It is hard to see how we could say the same, however, about biodiversity offsetting. Biodiversity is constituted partly by individual animals, who have morally relevant interests. At the very least, it seems implausible to suggest that biodiversity could be destroyed at impact sites without thwarting some of those interests.

Now, it is true that other animals could come into being at the offset site. Indeed, the idea that an 'equivalent' number of organisms will indeed emerge there is probably a core assumption of most offsetting schemes. But does their emergence at the offset site excuse what has happened at the impact site? A utilitarian might suggest that killing one set of animals and encouraging or facilitating the birth of another set of animals is a morally neutral conjunction of acts. According to what Singer (1979) calls the replaceability argument, while painful deaths are objectionable, there is nothing inherently wrong with killing animals, so long as an equivalent number of animals can come into being elsewhere, and go on to live healthy lives. However, it is difficult to see how any view that takes the rights of non-human animals seriously could endorse such a view. Those who hold deontological or hybrid views about rights and interests do not standardly believe that an actor can kill one person just so long as he brings another person into being. For Cochrane (2018),

---

[14] Barring, say, cases where the impact and offset sites are immediately adjacent, and all animals affected can simply move from one to another. Even here the idea that all harms can be avoided appears far-fetched. But in any case, impact and offset sites are not typically immediately adjacent: they could be in different parts of a country, and perhaps, if global offsetting schemes do emerge, in different countries entirely.

for example, animals have a *prima facie* right not to be killed, as well as a right not to be made to suffer. The problem with the utilitarian view is that it simply fails to recognize that animals are individuals whose lives matter *to them*. If each animal is the unique subject of a life, then showing, for example, a ewe proper respect means seeing her not as simply interchangeable, but rather as an individual end in herself (Korsgaard 2018). While carbon offsetting *might* be defended on the basis that it does not involve the commission of harm (a claim I have suggested in any case depends, implausibly, on ignoring other effects of emissions-related activities), no such claim can be made about biodiversity offsetting. To the contrary, it appears that defenders of biodiversity offsetting simply fail to take the interests, and rights, of individual animals sufficiently seriously.

This allows us to draw the following conclusions. In the majority of cases we are actually likely to confront, biodiversity offsetting is a practice that inevitably involves the commission of morally objectionable harms. The attempt to vindicate offsetting that is familiar from the carbon case—which suggests that harms do not arise in the first place, if emissions are offset—cannot get a grip in the biodiversity case, because biodiversity destruction at impact sites regularly and predictably causes harms to many animals. Moreover, if the introduction of offsetting makes it cheaper or easier to destroy biodiversity—when contrasted with alternative policies, such as the requirement to minimize or mitigate biodiversity loss—then offsetting policies facilitate such harms to animals. This is a problem that appears destined to apply even under highly idealized circumstances, and under any baseline, whether static, negative, or positive. Even if our baseline is demanding enough to preserve (or even augment) the total amount of biodiversity in existence, animals will likely be harmed, and their rights violated, in the process of offsetting.

Location, meaning, and global justice

The moral significance of location

One view about emissions holds that their geographical location is not by itself significant. By contrast, the location of biodiversity—and therefore biodiversity destruction—might well be significant. For one thing, location might be important in biogeographical terms: an identical slice of biodiversity might have one set of environmental effects if located in one place, but a different set of effects in another (Apostolopoulou and Adams 2017: 25).

In many cases, ecosystem processes in a particular place are sustained by specific assemblages of species. If conservationists were to remove part of an ecosystem and transplant it perfectly to another site, it could not be assumed that it would continue to perform those same processes, or at least not to the same degree. This adds further pressure to claims about the 'equivalence' of biodiversity at impact and offset sites. Perhaps equivalence needs to be judged not in terms of composition, but (instead, or also?) in terms of the number or quality of ecological relationships that given slices of biodiversity are likely to sustain.

For another thing—and this is my focus here—identical slices of biodiversity might sustain different *social* connections in different places. Even if people tend not to care about particular carbon molecules, they *do* often care about specific habitats, ecosystems, species, and plant and animal populations (Ives and Bekessy 2015: 571). One way of capturing this point is to turn to the language of attachment. People's plans, projects, identities, hopes, and fears are often bound up with their relationships with *particular* parts of the living world, rather than with generic types of habitat or ecosystem (Armstrong 2014). As a result, when people are excluded from impact sites, this can involve setbacks to important human interests. Indeed, we can say the same thing about cases where people are not actually excluded, but where those places are nevertheless transformed. With reference to climate change, Jennifer Szende (2022) argues that one way in which climatic change can undermine people's place-based attachments is by making spaces uninhabitable, but that another (and possibly much more widespread) problem is the transformation of places in such a way that they can no longer sustain the same identities and practices. As the indigenous scholar Kyle Whyte notes, some members of his own Anishinaabe community have suggested that if prevented from engaging in traditional activities (such as the harvesting of wild rice) in specific places, they would no longer recognize themselves as being part of the same communities (Whyte 2018: 132).

This kind of transformation of places will plausibly occur at many biodiversity impact sites. As Ives and Bekessy (2015: 572) put it, 'Creating offset sites away from where the biodiversity is being destroyed means that nature-based recreation and environmental education opportunities [at impact sites] are lost, natural amenities and environmental health are reduced, and places that shaped unique memories are markedly transformed'. When it comes to justifying offsetting, a critical question is whether these social costs are somehow made good when we facilitate the emergence of biodiversity in other places, bringing benefits to people located nearby. Here we can sound two cautionary notes. First, when drawing up offsetting schemes, policy makers simply do not insist that 'equivalent' social relationships must be sustained at offset sites. The quality of social relationships is just not a criterion for the evaluation of equivalence in the practice of biodiversity offsetting. As a result, we know relatively little about the human consequences of destruction and offsetting. Of course, it is perfectly possible that people at a restoration off-set site *could* come over time to develop important relationships with newly emergent biodiversity, but there is no guarantee that this will happen. People located nearby the new site might not care about any emerging biodiversity, or they may have other priorities. People might not even be able to access the offset site (Apostolopoulou and Adams 2017: 25), either because administrators are practising some kind of 'fortress conservation' or simply because it is far away from human habitation. Since social costs are not factored in when assessing the permissibility of offsets, we do not know if those costs are very high or very low. However, it seems plausible that in many cases they will be steep.

This first problem could in principle be alleviated by good institutional design. Policy makers could, for instance, make the flourishing of identities and attachments a desideratum of offsetting schemes, or even an additional index of 'equivalence'—although it is hard to envisage what this would look like in practice. Perhaps if one community loses an important attachment to an ecosystem, this is excused when another community of a similar size is able to develop a new attachment of equivalent intensity? But this brings us straight to the second problem. Even if 'equivalent' social relationships *did* come into being at the offset site—whatever that might mean—we might wrong people by disrupting existing relationships, and it is not obvious that this would be vindicated by creating opportunities for others to develop similar relationships. Respect for people's place-based interests—and an awareness of how badly things often go for the less advantaged in cases where local environments are transformed—might create quite a strong presumption against forcibly severing the link between people and local ecosystems. Anna Stilz, for example, argues that people possess occupancy rights that entitle them to remain in places to which they are attached and that others have a corollary duty not only to refrain from forcibly removing them from that space, but also 'not to interfere with one's use of the space in ways that undermine the shared social practices in which one is engaged' (Stilz 2017: 353). Even if we can imagine cases in which this duty is over-ridden—in cases, for instance, in which the basic rights of many people are at stake—showing due respect for people often means showing respect for the attachments that are central to their lives.

Offsetting, then, appears to be grounded in a picture of biodiversity as generic and interchangeable (at least within habitat types, and sometimes across types). But often we do not experience the living world as generic and interchangeable. As Nancy Fraser (2021: 119) puts it, 'The idea that a coal-belching factory here can be "offset" by a tree plantation there assumes a nature composed of fungible, commensurable units, whose place-specificity, qualitative traits and experiential meanings can be disregarded'. According to Apostolopoulou and Adams (2017: 23), offsetting 'reframes biodiversity as lacking locational specificity, ignoring broader dimensions of place and deepening a nature-culture and nature-society divide'. In the words of the Global Forest Coalition (2022: 4), finally, 'biodiversity offsets fundamentally ignore the reliance on biodiversity for local livelihoods, and de facto dismiss its local economic, social and cultural value'. But it also ignores the importance, for some communities, of communion with specific non-human others.

These concerns—about exclusion from or transformation of place—are most commonly voiced in relation to impact sites, given that it is these sites which are 'developed'. But similar concerns can also apply to offset sites. In the case of Rio Tinto's Madagascan offset, it is true that locals were excluded from the impact site where the mine was created. But locals were *also* excluded from the offset site. In fact, it was *because* Rio Tinto prevented swidden agriculture, which was (controversially) presented as being a major cause of biodiversity loss,[15] that the 'protected' site could be designated as an avoidance offset. If they had not prevented the practice of swidden at the offset site—and removed locals' only available means of subsistence in the area—Rio Tinto would not have been able to present their activities in Madagascar as biodiversity neutral. As one local put it, 'We used to cultivate manioc in the forest before this project came. Now we are not allowed anymore to plant in the forest . . . We now cannot grow enough food to feed our families' (Kill and Franchi 2016: 13). Promised payments that could have (however imperfectly) defrayed some of the opportunity costs involved have simply not materialized. As another local put it, 'they took our land and did not even compensate us. They said they would, but they never did' (ibid: 18). It is possible, of course, that this is an isolated case. But one study has found that 35 per cent of offset sites involve the displacement of people and negative effects on livelihoods (Sonter et al. 2018).

According to the Global Forest Coalition, making developers and subcontractors 'legally liable for delivering measurable biodiversity gains' provides

---

[15] As mentioned in Chapter 4, swidden is a form of itinerant agriculture, where locals clear vegetation from a site, grow crops for a year or more, and then move on. In Madagascar, swidden agriculture is known as *tavy*, and is mainly associated with growing rice and manioc.

'a clear incentive for so-called fortress conservation measures' at offset sites, leading developers 'to ban sustainable use or even access to the biodiversity [offset] site by women and other local inhabitants' (Global Forest Coalition 2022: 5). Such cases add insult to injury. In a lose–lose scenario for the local poor, the destruction of biodiversity at the mining site leads to exclusion, and the process of balancing out that destruction at the offset site *also* leads to exclusion. In that sense, biodiversity offsets can, at worst, represent 'a double landgrab, because a corporation will take away land from communities not only for the mine or the plantation or infrastructure development, but also for the area they plan to use for the biodiversity offset project' (ibid: 24).

What is the upshot of these arguments about the moral significance of location? They serve to place another nail in the coffin of the claim that biodiversity offsetting is a morally neutral practice and that biodiversity impacts that are successfully offset can be considered harmless. I have already argued that this claim fails in the case of the individual creatures likely to be harmed at impact sites. Here I have expanded the charge by showing that the interests of many human beings are likely to be set back in the process of offsetting, whether at impact sites, offset sites, or both. The existence of important place-based interests shows that the introduction of offsetting is not a morally cost-free process. Rather, when contrasted with strategies of avoidance and minimization of biodiversity loss, offsetting is likely to seriously affect the interests of many people who currently enjoy close and important relationships with particular tranches of biodiversity. Moreover, it is far from obvious that these moral costs should be seen as compensable when new relationships are facilitated elsewhere: if not, this is a problem that will affect even the best-designed offsetting policies. Though these problems are commonly marginalized in political discussions about offsetting and its merits, they ought to form a core part of our moral evaluation of the practice.

Opportunity costs and exploitation

Some scholars have welcomed carbon offsetting schemes because of their potentially redistributive effects. One consequence of investors in the global North buying carbon offsets from the global South is that money flows from the former to the latter (Hyams and Fawcett 2013: 92). Other things being equal, such flows can help to alleviate poverty in the South, and inequality between countries. In principle, we may be able to say precisely the same thing about the biodiversity case, *if* a truly global market in biodiversity offsets emerges over time. That market could also provide an avenue for substantial North–South financial flows. Indeed, even if practised exclusively at the national level, biodiversity offsets can still be said to have a redistributive effect. Domestic investors are likely to buy offsets in areas where the returns on land are lowest, on the assumption that this will be cheaper than paying to mitigate biodiversity destruction at impact sites. Other things being equal, this could serve to channel precious resources in the direction of poorer regions. The opening of an offset market might also make investment into a country more attractive for multinational corporations— inasmuch as it reduces the price of development—which could also have a redistributive effect, this time at the global level. Perhaps many people in the global South would rather conserve local ecosystems, rather than develop them. If so, offering them up as offset sites can help to dissolve the conservation/development dilemma, producing a secure income for people who are content to leave biodiversity undisturbed.

On further reflection, though, it is less clear whether the transfers involved would really advance global justice all things considered. We should consider trades not simply in terms of a transfer of resources from rich parties to poorer parties, but in more holistic terms, taking into account the broader opportunities that the rich gain and that the poor give up. It is clear that for better-off actors in the global North, the availability of offsetting reduces the price of development: if it did not, they would turn to other strategies such as avoidance or minimization. But it is less clear that the sale of offsets should be considered a win overall for poorer actors. Selling offsets will certainly bring in income, which may be very welcome in the short term. But it might reduce opportunities to escape from poverty or inequality in the medium or longer term, entrenching rather than reducing North–South inequality.

One way of capturing this worry is to return to the argument about opportunity cost baselines from Chapter 4. The vagaries of demand and supply could lead to any of a number of different market outcomes. If the supply of offset sites was limited, but demand was high, then it is perfectly possible that those who offer offsets up to the market could extract revenues in excess of what they would ordinarily have earned (assuming that this is still below the cost, to developers, of alternative mitigation strategies). By contrast, if there were lots of sellers in an offset market, the price of biodiversity offsets appears much more likely to settle around what would be expected under a status quo or willingness-to-accept baseline. This would involve actors comparing offers for their land with the income they were otherwise likely to have earned, and accepting offers broadly equivalent to those lost earnings. In practice, offsetting markets appear to have settled closer to our second scenario, with payments reflecting what would have been earned with land under the status quo within the global economy. And indeed, that is what international standard-setting institutions suggest *should* happen. The International Finance Corporation (IFC), for instance, sets ethical standards for its own activities, and also maintains a set of principles (the 'Equator Principles') for other financial institutions that invest in low-income countries. The IFC's standards clearly connect offsetting payments to the (status quo-defined) opportunity costs incurred by those who live in offset sites (Bidaud et al. 2018). The assumption that prices should reflect opportunity costs plus any administrative costs (and any restoration costs, in the case of restoration offsets) appears to be widespread in offsetting practice (see Koh et al. 2019, Simpson et al. 2021).

If payments for offsets do indeed coalesce around a status quo baseline, then the same worries that I expressed about conservation burden sharing in Chapter 4 reappear. In the context of a highly unjust global economy, there is no reason to treat the level of income locals *would have* earned in a given place as a morally privileged benchmark for assessing fair transactions. Quite the contrary. When outsiders take advantage of the weak structural position in which some actors find themselves, in order to deny them a fair return on their sacrifices, then those transactions can be viewed as exploitative in nature. Those concerned about global justice, of course, need then to flesh out an account of what would constitute a fair distribution of opportunities broadly considered. An anti-poverty baseline would impugn offsetting payments that leave locals below a reasonable poverty line. An egalitarian baseline, by contrast, would connect fair payments with what actors could have earned if their opportunities had been broadly equal. With either model, it appears likely that many offsetting transactions should be considered as exploitative in character. And if developing their land was the only realistic path out of disadvantage for many people in the global South, then offsetting schemes that merely seek not to worsen their existing standard of living need not be considered distributively just.

Conclusion

This chapter considers a variety of worries about biodiversity offsetting, ranging from contingent worries that could in principle be alleviated by good policy design, through to concerns that appear destined to apply to any instance of biodiversity offsetting, however well intentioned. Contingent worries are not, of course, less significant for being contingent: it might be perfectly *possible* to design out problems, but very unlikely that policy makers will in fact do so. For that reason, investigating the *typical* effects of biodiversity offsetting is important. The fact that offsetting schemes frequently fail to achieve their core objectives (Buller 2022: 253), and that even when successful, the fruits of restoration policies may take centuries to accrue (Curran et al. 2014), is information that needs to be considered when evaluating the policy. It is also important to reflect on the effects of offsetting in a suitably holistic sense. Offsetting policies tend to be introduced against a legal and political backdrop that already includes some constraints on the destruction of biodiversity. Frequently, however, the introduction of offsetting accompanies—and justifies—a weakening of those constraints. Offsetting forms part of a larger trend towards making conservation more 'business friendly', in which the introduction of markets has come to replace tighter direct regulation of environmental impacts (Kill and Franchi 2016: 19). Research indicates that in jurisdictions where offsetting is available, other environmental protections tend to be weaker (Ives and Bekessy 2015, Phalan et al. 2018, zu Ermgassen et al. 2019b). This suggests that once offsetting enters the picture, the mitigation hierarchy is honoured more in word than in deed. Meanwhile, a failure to communicate to the public the implications of the widespread adoption of negative baselines for biodiversity loss means that the consequences of offsetting (and of 'no net loss' policies on biodiversity) are not well understood. There is a real danger that the embrace of offsetting has helped to sow a mistaken belief that the problem of biodiversity loss is under control when it very definitely is not.

Still, it is true that, in principle, it should be possible to mitigate some of the pathologies associated with offsetting. As discussed, in some places offsetting is associated with serious injustices, including forcible exclusion from offset sites, and/or the prohibition of subsistence activities. In theory, it could be made a condition of offsetting that people are not excluded from, or robbed of, their livelihoods, whether at impact or offset sites. I also argue that offsetting transactions can be exploitative and can make the long-term prospects of poor people in the global South worse rather than better. But in principle, policy makers could adopt an egalitarian baseline that obviates that worry. Finally, policy makers could adopt static or positive baselines for biodiversity loss, and treat avoidance offsets with caution, thereby ensuring that biodiversity is not offset to destruction.

Such policies are somewhat unlikely, but might they vindicate the claim that offsetting could potentially be a morally cost-free process, at least in principle? The answer is no—because, as I argue, there are some worries about offsetting that appear destined to plague any policy. What I call the harm-cancelling argument fails in the case of biodiversity (in fact, I suggest that it fails in the case of carbon offsetting, too, when we assess that practice in a suitably holistic way). Offsetting appears certain to license serious harms to individual animals, and if we are prepared to take seriously the thought that at least some of those animals have rights, then we must recognize that destruction at impact sites is highly likely to violate those rights. At the very least, we must consider this in discussions of the policy. Moreover, the conversion of impact sites will often set back important human attachments to biodiversity and to place, committing harms not obviously wiped clean by creating possibilities for people to engage with biodiversity elsewhere. Even if very well designed, biodiversity offsetting can involve serious moral costs, both to humans and to other animals. Everyday offsetting practice involves very serious moral costs; but even the most ideal offsetting policy ought to be treated with considerable caution.

6
Half Earth and beyond

As it has become ever more apparent that existing conservation policies are failing to arrest the loss of biodiversity, academics, practitioners, and policy makers have called for responses that are more ambitious, synoptic, and transformative in character. The need for such responses has been explicitly recognized, for instance, in publications by the Intergovernmental Science-Policy Platform on Biodiversity and Ecosystem Services (IPBES 2019), and in the recent Kunming–Montreal Global Biodiversity Framework (GBF). In the scholarly literature, it is by now almost unusual *not* to apply the word 'transformative' to discussions of biodiversity governance (Wyborn et al. 2020b, Massarella et al. 2021). But what would genuinely transformative conservation policies look like? In this chapter I examine perhaps the most ambitious conservation policies ever proposed, which emanate from the 'Half Earth' movement. In a nutshell, Half Earth calls for the creation of very large protected areas, which will be insulated from further economic 'development'; when combined, these areas should cover fully half of the surface of the earth. On land, this would mean a total area four times the size of Russia. As such, the claim that the Half Earth proposal represents 'the largest land governance proposal in history' (Ellis and Mehrabi 2019: 22) comes with little danger of overstatement. At sea, meanwhile, it would mean an area larger than the Pacific Ocean.

In this chapter, I use 'Half Earth' as an umbrella term for proposals variously entitled Half Earth, the Global Deal for Nature, and Nature Needs Half.[1] There are some differences between these proposals: the Global Deal for Nature, for instance, is a hybrid proposal aiming to protect biodiversity and also sequester carbon. But these differences aside, there is what Dinerstein et al. (2017: 535) call a 'striking example of convergence' around the goal of protecting 50 per cent of the surface area of the globe. On their view, the model provides what conservation efforts to date have been sorely lacking: 'an endgame' (ibid: 542). Half Earth supporters claim that putting the

---

[1] Each maintains a distinct organizational presence, as can be seen on their websites: half-earthproject.org, natureneedshalf.org, and globaldealfornature.org.

Global Justice and the Biodiversity Crisis. Chris Armstrong, Oxford University Press.
© Chris Armstrong (2024). DOI: 10.1093/9780191888090.003.0007

model into practice would allow us to maintain a healthy biosphere, avoid mass extinction, and preserve the ecological processes that sustain human societies. In fact, such policies may be *necessary* for these goals to be met, rather than merely helpful: in the words of its most famous exponent, Edward Wilson (2016: 3), 'Only by setting aside half of the planet in reserve, or more, can we save the living part of the environment and achieve the stabilisation required for our own survival'.

Six features of Half Earth proposals are worth briefly noting at the outset. First, Half Earth is a serious proposal, with many famous advocates, with dedicated institutional presences, and it appears to be influencing conservation practice already. The website of the Global Deal for Nature, for example, suggests that the plan has been endorsed by people from 92 countries, and by an impressive array of conservation organizations. Notwithstanding the world's failures to meet past conservation goals, the GBF recently embraced the target of ensuring that 30 per cent of terrestrial and marine areas are protected by the year 2030. It would not be surprising if the world's leaders go on to adopt the goal of 50 per cent protection by 2050. If they do so, Half Earth advocates will have played a significant role in preparing the ground.

Second, Half Earth is an *area-based* conservation programme and aims to preserve half of terrestrial and marine ecosystems by surface area in the hope that this will then secure the future of biodiversity. Half Earth is in the first instance a target, then, for area protection, rather than for the amount of biodiversity that is preserved. In fact, because of its geographical concentration, supporters expect that the amount of biodiversity preserved by Half Earth policies would be much greater than fifty per cent. Wilson, for example, argues that Half Earth would enable the preservation of 85 per cent of the planet's surviving species (Wilson 2016: 186).

Third, Half Earth supporters aim to preserve both terrestrial and marine areas. But they do not see land and ocean ecosystems as fungible when it comes to meeting that goal. If they did, the Half Earth proposal would be far easier to meet, and the challenges I discuss later would bite much less keenly. Elsewhere I argue that 80 per cent of the ocean should be strongly protected from destructive economic activities; I also argue that policy makers could achieve this without interfering greatly with people's nutritional security, employment, or access to the ocean (Armstrong 2022). If such a goal was achieved, policy makers would already have protected more than half of the earth's surface (and over ninety per cent of its habitable volume), with the implication that Half Earth could be satisfied without protecting any land at all. Half Earth, though, aims to preserve half of both land *and* sea. This is a much more demanding target.

Fourth, Half Earth aims to preserve relatively large and interconnected areas of land and sea. There are several reasons for preferring protected areas to be larger rather than smaller in size. It is suggested that larger areas will be more resilient to long-term environmental change (Noss et al. 2012); for instance, the ability of wildlife to move in response to changing conditions is an important buffer against environmental challenges and for that reason larger size correlates negatively with extinction risk (Purvis et al. 2000). Furthermore, members of some species (such as large predators) must roam over large areas in order for populations to be viable (Pimm et al. 2018). As well as being large, protected areas would also ideally be inter-linked, with corridors allowing animals to roam across regions. This would facilitate ecosystems' adaptation to climate change and other pressures, and allow the interchange of genes (Noss et al. 2012, Dinerstein et al. 2017). Other things being equal, protecting larger areas will mean that more, rather than less, biodiversity is conserved. This would be good for the animals who partly constitute the ecosystems in question. Further, since the robustness and vitality of various ecosystem functions correlate positively with biological variation (see Chapter 2), it would be good for humans, too.

Fifth, the practical focus of Half Earth will fall disproportionately on tropical and subtropical regions. Officially, Half Earth aims at the protection of 'biodiversity hotspots', which can be identified on the basis that they contain large numbers of endemic species (that is, species found nowhere else on earth). In principle, hotspots could be found anywhere. In practice, they are highly skewed towards the tropics. This is partly because so much biodiversity has been destroyed elsewhere. But the skew is also explained by the latitudinal biodiversity gradient, which tells us (albeit for somewhat contested reasons) that marine and terrestrial ecosystems are much more productive and complex the closer they are to the equator (Dowle et al. 2013). The tropical focus of Half Earth is not always apparent at first glance. Locke (2014: 365), for instance, suggests that protected areas should aim to cover all distinctive ecosystem types (or 'biomes'), each of which can be broken down into a large number of ecoregions. Dinerstein et al. (2017) describe fourteen terrestrial biomes, including seven varieties of forest (tropical and subtropical moist broadleaf; tropical dry broadleaf; tropical coniferous; temperate broadleaf; temperate conifer; boreal; and mangrove), and seven non-forested biomes (tropical grasslands; temperate grasslands; flooded grasslands; montane; tundra; Mediterranean forests, woodlands and scrub; and deserts and xeric shrublands). But when they go on to identify which ecoregions could feasibly achieve 50 per cent protection, it is clear that tropical and subtropical forest ecosystems are the most heavily represented category by a very great distance (ibid: 537). Wilson, by contrast, is upfront about the fact that, given their exceptional levels of extant biodiversity, most of the ecoregions selected for protection will be situated in the tropical parts of the world (Wilson 2016, chapter 15). From the perspective of those who control the budgets, protecting tropical hotspots looks like an 'efficient' use of conservation resources. Protecting biodiversity outside of these hotspots, by contrast, can be expected to yield more meagre results for the same resources.

Sixth, note that, given the present state of the world's terrestrial ecosystems, implementing Half Earth would require a halt to ongoing habitat loss, but also some degree of restoration or rewilding. In the ocean, some ecosystems (such as mangroves and coral reefs) are under serious pressure, but more than half of marine habitat is still extant. But back on dry land, cropland, grazing, and commercial forestry, plus human settlements and infrastructure, already take up a little over half of the world's land area (Ellis and Mehrabi 2019: 25). As such, Half Earth would necessarily involve giving up some land currently under cultivation (as discussed later).

The question I pursue in this chapter is whether Half Earth proposals are compatible with global justice. Half Earth advocates are right that existing conservation policies—which are largely incremental and piecemeal in nature—have failed to arrest or reverse the biodiversity crisis (IPBES 2019). They are also right that this crisis threatens the most basic interests of present and future people, as well as those of other animals we share the planet with. In that context, the ambition of Half Earth is to be commended. Nevertheless, I suggest that there are grounds for serious caution about these proposals. At the very least, they suggest that significant side policies would need to be in place if Half Earth is to avoid exacerbating global injustice. But the sternest challenge for these proposals might be a comparative one. If alternative policies are available that would be similarly effective in staving off biodiversity loss but would not have such serious negative implications for global justice, then this fact would cast serious doubt on the permissibility of the Half Earth project. It is for that reason, I suggest, that Half Earth must ultimately be rejected.

The pendulum of conservation politics

To really get to grips with Half Earth proposals, it is vital to situate them within long-standing debates about the priorities of conservation policy. These have taken a somewhat dialectical form. First, chronologically speaking, we can identify a 'traditional' conservation model, which focused on

establishing large protected areas or nature reserves. In the US, the archety-
pal objective of this approach was the creation of the large 'wilderness park'
with minimal human presence, on what is sometimes called the 'Yellowstone model' (Marris 2011). Notoriously, the establishment of Yellowstone
National Park in 1872 involved the displacement of indigenous communities from their traditional territory, and onto separate reserves. Almost two
decades later, the establishment of Yosemite National Park was also accompanied by the 'premeditated, careful, and callous eviction of Mimok, Yokut,
Paiute, and Ahwahneechee peoples' (Murdock 2021: 239).

But the general approach has a longer and wider history. In Britain during
the 1700s and 1800s, landowners on many aristocratic estates progressively
'cleared' poor residents in order to secure space for charismatic wildlife (as
well as for hunting, which was often presented, however dubiously, as a
pro-conservation activity). Much the same model was then exported to the
colonies, where it influenced conservation policy for many decades. An influential 1931 report entitled *Proposed British National Parks for Africa*, for
example, called for the creation of 'apartheid parks', which would be freed
from the burden of (African) human impact, and administered by white colonizers (Adam 2014: 35). While I call this the traditional model, it is also
important to recognize that its traces have survived the formal retreat of colonialism (ibid: 34). Many contemporary national parks in both Africa and Asia
share a footprint with hunting reserves set up by and for colonizers (Martin
2017: 21). Their governance can still be highly exclusionary, with uneven
impact sometimes felt along racial lines. In Kenya, for example, the government abolished long-recognized hunting rights of the Maasai people when
it established the Amboseli National Park (Murdock 2021: 243). After the
establishment of the Mkomazi Reserve in Tanzania, meanwhile, the memories of evictees were figuratively 'erased by representations of its landscape
as a wilderness restored' (Brockington and Igoe 2006: 450). There has been
considerable continuity in the way that governments in both the colonial
and post-colonial periods have undermined the claims of people traditionally
dependent on forests for subsistence (Kashwan 2017: 33).

It should be relatively clear what is objectionable about this first model. It
embodies a stark political hierarchy between those who are entitled to determine conservation policy and those destined to be objects of it. Conservation
politics is characterized here by what republicans call domination, involving the exercise of arbitrary power over people who are considered to be
rule takers but not rule makers (Pettit 1997). The burdens of conservation
activities are also spread highly unevenly, to say the least. Millions of people's lives have been disrupted, as traditional subsistence activities are newly

outlawed, often on spurious grounds. The connection between people and place is brutally severed. The benefits of conservation, by contrast, typically flow to the privileged. In Africa and Asia under colonialism, a story familiar from European history was rerun, in which conservation policies drove what Marx called 'primitive accumulation'—wresting people from the land through violent enclosure, and thereby creating a floating labour-force destined for the cash economy (Büscher and Fletcher 2020: 74). So-called Game Laws removed the hunting rights of poorer communities and allowed privileged elites to monopolize such rights. In Europe, according to Perelman (2007: 53):

> over time the Game Laws reflected an emerging hegemony of property relations in which the interests of capital and the gentry coincided. The gentry could enjoy the prestige of hunting, while the capitalists could profit from the labour of people who were forbidden to hunt as a means of subsistence.

In Africa and Asia, this economic hierarchy was often intertwined with a racialized one.

More recently, the 1990s saw the identification of a new wave of 'fortress conservation'. According to Dominguez and Luoma (2020), fortress conservation is characterized by the creation of protected areas from which locals are excluded; by the use of perimeter patrols by armed park rangers; and by the restriction of activities inside those perimeters to tourism, safari hunting, and scientific research. At the extreme, conservation has been seen as incompatible with any significant human presence (with exceptions made for the self-designated—and often white and wealthy—stewards of biodiversity). According to one critic:

> global conservation as we know it today is deeply reliant on earlier Euro-Western forms of land-use management and wilderness definitions that are [. . .] openly hostile and violent toward the presence, land-use philosophies, environmental ethics, cosmologies, and environmental conceptions of people of color and subordinated groups (especially indigenous peoples) more specifically (Murdock 2021: 236).

An especially worrisome development, during the 1990s, was the emergence of militarized forms of biodiversity protection in the global South, in some cases funded by international non-governmental organisations (Duffy 2014), and the displacement of 'conservation refugees' (Dowie 2009). More recently, Payment for Ecosystem Services and REDD+ models have also prompted widespread concerns about the displacement of locals, including indigenous communities, from the land and from traditional subsistence activities. These effects may not have been intended by the designers of such funding models. Nevertheless, 'almost by default, and often against the wishes of project designers, "fortress" forms of conservation forestry in reserves, or uniform plantations, under clear state or private control, became the only way that carbon value can be appropriated through these mechanisms' (Leach and Scoones 2013: 965).

The second model, which competed with and partly came to displace the first model from the 1990s onwards, has been called the 'New Conservation'. The weak performance of fortress models of conservation, 'plagued by protests and sabotage by local people who view them as an illegitimate and oppressive imposition' (Siurua 2006: 75), prompted reflection on whether a different approach was necessary. The New Conservation represents a reaction against the traditional/fortress model, with the pendulum swinging in the direction of an avowedly 'people-friendly' approach (Martin 2017: 4). Prominent New Conservationists have argued that we ought to abandon the ill-conceived focus on preserving wilderness and undisturbed nature, which in any case 'never existed, at least in the last thousand years' (Marvier et al. 2012). Early US conservationists like John Muir, for example, improperly downplayed the extent to which North America's ecosystems had been shaped significantly over time by generations of indigenous inhabitants. Their vision of a pristine wilderness was in that sense 'produced', and often violently imposed on locals (see e.g. Guha 1989, Neumann 1998, Whyte 2018). Rejecting that picture, the New Conservationist favours a model in which people and non-human life mix freely, rather than being separated into distinct 'human' and 'natural' zones. The New Conservationist has often embraced social objectives, such as poverty alleviation and political inclusion, alongside the environmental goal of biodiversity protection. By making conservation work for people, supporters suggest, we make conservation more likely to actually occur, as well as fairer. Making conservation work for people will in turn involve identifying and tapping into the many economic interests served by effective conservation—and this means accepting that business and conservation can be friends rather than enemies.

The New Conservation has come in for trenchant criticisms of its own, however (see e.g. Miller et al. 2014). It has been criticized for demonstrating a thoroughgoing anthropocentrism, manifested in its (apparent) belief that nature must be good *for people* in order to have value (see e.g. Batavia and Nelson 2017). That criticism is not necessarily a fair one: some New Conservationists have explicitly accepted that nature has intrinsic value (see e.g. Marvier 2014: 2). What is crucial to their view is the strategic claim that (human) social objectives require renewed emphasis if the activity of conservation is going to achieve its objectives: *effective* conservation will be people centred, to a large extent. More worryingly, perhaps, the approach is often criticized for leaving protection to the market, and especially to corporate interests (Doak et al. 2015). As one defender of the New Conservation has put it:

> If species or habitats are to be conserved they must not be protected from market forces as that will place them in the hands of an inefficient state that will allow them to degrade [. . .] Rather, they must be fully exposed to a market [. . .] where their uniqueness and scarcity will lead to high economic values being placed on them (Hulme and Murphree 1999: 280).

The New Conservationist intervention has been linked to a wider collapse in support for state direction (and redistribution), and a resurgence of faith in the market to deliver social objectives instead. Both are characteristic of neoliberalism, with which the New Conservation is thought by its critics to be on close terms (see Holmes et al. 2017). Both, it is said, share the Promethean belief that never-ending economic development is possible, without living systems being degraded beyond repair. But to date, there is little evidence that any absolute uncoupling of growth and biodiversity destruction has been forthcoming.

Finally, we can observe a more recent counter-movement, which has been called 'neoprotectionist' in character (Büscher and Fletcher 2020). This third model emphasizes that even if there is little (or even no) pristine wilderness in existence, there are still great swathes of 'wildness', which are home to a huge amount of biodiversity. Integrating 'people and nature' has meant, critics claim, subsuming nature to the impulses of the market. Given the scale of the biodiversity crisis, and given the lamentable failure of extant conservation policies to arrest it, another approach is necessary. It is in this third movement that we can situate the Half Earth project (see Locke 2014, Wilson 2016, Cafaro et al. 2017, Dinerstein et al. 2017, Kopnina et al. 2018). Half Earth counsels a pendulum swing back to the idea of large, strongly protected regions. The land that comes to be protected might comprise strict nature reserves, national parks, and other kinds of managed area, all of which are frequently grouped under the general collective term 'protected areas'. But can Half Earth policies be justified? And what might its moral costs be? These are the questions this chapter addresses.

Moral costs

In this chapter I do not seek to assess the scientific basis of Half Earth proposals. Wilson (2016), for instance, suggests that protecting 50 per cent of land and ocean will allow most species to survive; according to some critics, such claims remain uncertain, given our limited knowledge of the ecological and human processes involved in determining biodiversity outcomes. As I noted in the Introduction to this book, Wilhere (2021) argues that we would require an 'IPCC-like' scientific effort to pool such knowledge before Half Earth proposals could be properly evaluated. Instead, I try to inform debates about the pros and cons of Half Earth proposals by seeking to identify the moral costs—to humans—that could ensue if the idea was adopted. Though the questions of global justice I raise are not the entire story when we come to evaluate Half Earth, they are an important part of the story.

Exclusion

Dedicating one half of the land to 'nature' and one half to humanity might be taken to mean that humans should be excluded from the former.[2] If so, it seems that Half Earth will involve the return of fortress conservation on a grand scale. The worries about that possibility should be obvious. Many people have close connections to particular places, including biodiversity hotspots. Severing those connections should be seen as a very high moral cost for conservationists to incur. One kind of cost would be the loss of livelihood, in cases where people were engaged in subsistence harvesting activities, say, within a newly designated protected area. Another would be the loss of connection to particular places that play an important role in sustaining people's identities and world views. In principle, people required to desist from harvesting activities could be offered other economic options instead, and people displaced from particular places could be offered alternative places to live. But reflection on real-world cases of displacement tells us that people dislocated in this way suffer profound setbacks to their well-being, for which the substitution of other livelihood options, or relocation even to an ostensibly similar place, is typically a poor remedy (Whyte 2018). Relocation is frequently

---

[2] Alternatively, it might simply mean that some economic activities should be forbidden. But note that this too could be seen as a form of displacement. On Brockington and Igoe's (2006: 425) influential account, conservation displacement comprises both 'forced removal' and 'economic displacement', since preventing people from engaging in existing forms of livelihood will often mean they are obliged to leave for elsewhere.

experienced as an assault on displaced people's agency, and proof that their connection to the land has not been acknowledged or respected. The outcomes for displaced people in terms of mental and physical well-being are often poor, and their dislocated status can expose them to exploitation and abuse.

It is important to emphasize, moreover, that if Half Earth did indeed have these exclusionary implications, they would very likely be highly racialized. Reflecting on the years leading up to the emergence of the Convention on Biological Diversity, Jessica Dempsey (2016: 58) identifies a conservation politics that very much 'has its gaze fixed upon the lands of the Global South'. If Half Earth policies skewed protection towards the tropics and subtropics, this could lead to the exclusion of modest numbers of white people, but very large numbers of people of black, brown, Asian, and Latin descent. In so doing they would replay, in a somewhat novel register, a racial politics of conservation that is all too familiar from the 'traditional' conservation model. They would also disproportionately impact indigenous people, given that the land on which indigenous people live is much more likely to have high levels of intact biodiversity (IPBES 2019). We might even say that coercive and exclusionary policies would punish indigenous people for their success in preserving biodiversity: countries that had already destroyed most local biodiversity would not, on this basis, be an especial target of such efforts, but those that had not would clearly be in the sights of conservation administrators. From the viewpoint of global justice, coercive exclusion would be an exceptionally high price to pay for biodiversity protection. In this context, the suggestion that 'Those of us whose wealth and privilege are based on the ongoing plunder of nature within an imperialist hierarchy . . . should consider the implications of trying to legitimize further expropriation with appeals to saving nature' (Napoletano and Clark 2020: 42) is highly salutary. To date, however, there has been an unfortunate dearth of scholarly engagement with the possible racial and colonial dimensions of Half Earth proposals.

How vulnerable Half Earth is to these very serious worries is not wholly clear. This is because its proponents have been 'ambiguous about the exact forms and locations of the new conserved areas being called for' (Schleicher et al. 2019: 1094). On the basis of Dinerstein et al.'s proposal, one estimate suggests that as many as 170 million people currently reside in places likely to be earmarked for protection as part of nature's half (Schleicher et al. 2019: 1095). Would these people be excluded from those places? Half Earth proponents have not been as transparent in answering this question as many would like. Certainly some argue that there should be a mix of different categories of protected areas—from strict nature reserves, to managed areas, to parks (Kopnina et al. 2018). In at least some of these, some human presence could be permitted (after all, as Schleicher et al. note, around a quarter of a billion people already live within protected areas). Wilson (2016: 3), by contrast, did appear to view nature's half as a network of huge reserves, to be protected from human impact. In a more recent piece, several Half Earth scholars have engaged with prominent criticisms of the proposal, without quite settling the question of whether people are to be excluded from protected areas or not. Crist et al. (2021: 3) are clear that 'large-scale nature protection will bar corporate access from much of the world'. But they do not spell out what their proposals mean for people who currently live in areas that are to be designated as protected. They agree, for example, that it is 'crucial to partner with Indigenous and local communities near protected areas' (ibid 2021: 3); but the choice of the word 'near', as opposed to 'in', is ominous.

In general, it is fair to say that defenders of Half Earth have not given enough explicit attention to how any transition towards the protection of nature's half could be justly achieved. Their focus has largely been on the goal, rather than the process by which it might be secured, or the human costs that might result from pursuing it. But the normative obstacles in the way of any suggestion that humans should be removed from nature's half are very considerable. Wilson, for instance, might have believed that a policy of displacement could be justified all things considered, given the scale of ecological problems we now face. Displacement is bad, he might have said, but mass extinction is worse, and so are the consequences for humans of a widespread collapse of the earth's core ecosystem functions. Such a claim could underpin a kind of 'last resort' justification for expulsion. But any such argument would face two serious problems. First, it will rarely be clear that the ecosystems in any *particular* place must be preserved lest we hurtle towards ecological destruction (this is only partly because the consequences of destroying the ecosystems in any particular place will be hard to predict). Second, it seems unlikely that the strategy of displacement and protection *will* be our only option, even when protecting the most basic ecosystem functions is at stake. What if, for instance, the reason half of the earth needs to be protected is because many people in the 'human' half have refused to modify their consumption habits (as discussed later), even though many of them could easily do so? Trying to justify displacement on the basis that human survival depends on it is one thing. Trying to justify it on the basis that people have refused to take *other* steps that would ameliorate threats to biodiversity is quite another. The problem with many last resorts is that they turn out not to be last resorts at all.

Crowding

The exclusion objection focuses on where people are moved *from*, thereby severing a connection between people and place. The crowding objection focuses on where they are moved *to*, and points to the human consequences of squeezing people into half of the earth's land space. Büscher and Fletcher (2020: 36) conjure up a dystopian vision of 'a population herded into urban areas' as a result of the designation of protected areas within half of the earth's surface. This reprises their earlier claim that Half Earth would entail 'forcibly herding' people into 'increasingly crowded urban areas' (Büscher and Fletcher 2016). Such crowding, we are to presume, would be unpleasant, unhealthy, and probably inegalitarian.

This challenge is much harder to support, however. More than half of the global population currently lives within urban areas, which only take up 1–2 per cent of the world's land (Ellis and Mehrabi 2019, Prendergast and Vettese 2021). In that sense, it is far from obvious why the entire human population could not live very comfortably within 50 per cent of the land without being unduly crowded. On the one hand, urban living combined with high-quality public good provision and access to healthy green and blue spaces can be both fulfilling and environmentally friendly (Voytenko et al. 2016). On the other hand, there is no obvious reason why restricting humans to half of the land's surface would require them to live in urban settings if they did not wish to in any case. In short, the concern about crowding seems to add little to the first (valid) concern about exclusion. Even more regrettably, the focus on urban crowding misdirects our attention away from some of the core conflicts that are likely to be prompted by Half Earth proposals. The world *is* likely to see increasing pressure for land if policy makers embrace Half Earth—indeed this will be true whether they embrace Half Earth or not. But it is not the desire for space for habitation that is the key driver of land scarcity. By far the greatest source of pressure for land comes from animal agriculture (to which we can add important additional demands, as we try to mitigate our way out of a looming climate catastrophe: the demand for land for wind and solar harvesting and for land capable of afforestation or other forms of carbon capture). The increasing demand for land in agriculture is driven both by a growing population, and by changing consumption habits—and in particular the much more widespread adoption of meat-based diets. Twenty-seven per cent of the world's landmass—an area the size of the Americas—is now given over to pasture and feed-crops for meat and dairy production (Prendergast and Vettese 2021: 19). It is here, rather than in habitation, that we find the real pinch-point for Half Earth. Reserving half of the earth for nature might not require urban crowding. But it might well be impossible to achieve without the much more widespread adoption of vegan or near-vegan diets.

Deprivation

It is vital that the burdens of conservation policies are fairly shared, and this means, among other things, that those burdens should not be placed on the shoulders of those who are very badly off, if that is at all avoidable (Armstrong 2019b). Asking the worst off to bear conservation burdens represents bad policy in at least two ways. On the one hand, since the worst-off *cannot* carry the burdens of conservation policies, asking them to do so is a recipe for policy failure. On the other hand, it would rarely be fair for the worst off to carry conservation burdens in any case, since any contributory responsibility they might be thought to bear for causing environmental damage is likely to be vitiated by their straitened circumstances, and others possess much greater ability to pay. As I argue in Chapter 3, people who are *not* very badly off, meanwhile, should bear burdens in relation to their contribution to the problem, first, and their capacity, second. If there were no other way of avoiding the hugely negative consequences that would arise from unmitigated biodiversity loss, then we might be justified in requiring the worst off to bear a share of the burdens of tackling those problems. But when there *are* accessible options that do not involve the worst off bearing those burdens, requiring them to do so is unjust.

It seems likely, though, that Half Earth—conceived of as a large-scale mitigation strategy for the biodiversity crisis—would, at least in the absence of robust side policies, misallocate conservation burdens, generating significant costs for some of the world's most disadvantaged people. In the contemporary world, as we have seen, conservation burdens are already pervasively misallocated, with costs overwhelmingly falling on the shoulders of those who can least afford to bear them. International efforts to spread conservation burdens more equitably are very seriously underfunded (McCarthy et al. 2012). But Half Earth would triple the area protected by global conservation policies (Wilhere 2021: 2). If it was not accompanied by substantial and well-funded side policies, it would likely compound existing injustices. This in turn would probably lead to backlash effects and popular resistance to Half Earth.

One major worry is that by reducing the land available for farming, Half Earth might reduce the number of calories available to people overall, both in particular ecoregions and on the global scale. According to Mehrabi et al. (2018: 409), Half Earth would lead to a loss of 11 per cent of food calories produced globally, and 29 per cent of calories in some regions. This could lead to absolute deprivation in many places, as well as deepening existing deprivation. Globally, it could be expected to lead to a spike in food prices, not least since land prices can be expected to rise in a non-linear fashion as supply diminishes (Ellis and Mehrabi 2019: 27). Considered alone, this could pose a major challenge to the food security of the poor. Considered alongside competing claims for land generated by various climate mitigation strategies, we could envisage a perfect storm of rising food prices.

All of this implies that Half Earth would, unless accompanied by robust side policies, impact very negatively on the nutritional security of the poor. As mentioned, however, it is possible to push back somewhat on the worry about food prices and demand for land. The precise form that any trade-off between biodiversity conservation and eradicating hunger takes is not pre-determined; rather, it is conditioned by the choices that political and economic actors make. Most studies suggesting a trade-off between biodiversity conservation and nutritional security have focused on total food production, rather than investigating the mechanisms that actually affect the food security of the poor (Fischer et al. 2017). To date, though, real-world food insecurity has not been driven by shortfalls in total food production, but by its inequitable distribution. This alerts us to the possibility that any negative nutritional consequences for the disadvantaged would be mediated by political and economic structures, rather than being an inevitable feature of a Half Earth world. For instance, take the food choices of the advantaged. Mehrabi et al. (2018: 410) suggest there is a conflict between Half Earth proposals and the food requirements of humanity. But their model assumes that 36 per cent of global crop calories will be fed to livestock. Half Earth might, *given existing food preferences and inequalities in purchasing power*, lead to waves of nutritional deficiency. But it is less clear that the same outcome would follow if much of the world reduced its meat consumption. Alternatively, the threat Half Earth poses to the food security of the poor could be mitigated by way of robust redistributive policies seeking to ease poverty at source. All of this suggests that, for Half Earth to be pursued justly, it could not be pursued in isolation from policies seeking to change the priorities of the global food system nor from the broader distribution of resources and opportunities.

Exploitation

Conservation has many costs, including opportunity costs. Do advocates of Half Earth believe that people whose livelihoods are affected by the Half

Earth project are owed some kind of assistance or compensation? The answer to that question is not entirely clear. Wilson, for instance, does not appear to have ever addressed the question. If locals are to be excluded without any form of assistance, then the kind of exclusion that could attend Half Earth is even more worrisome. Even if we think that material assistance could never really make up for the plight of indigenous people excluded from their homes and life-worlds, to not even address the question is seriously remiss. But perhaps some Half Earth advocates do believe that opportunity costs, for instance, should be shared fairly. Strikingly, many influential contributions (e.g. Locke 2014, Kopnina 2016, Kopnina et al. 2018, Crist et al. 2021) simply do not address the financial cost of implementing Half Earth at all. Dinerstein et al., by contrast, do make brief reference to the issue, suggesting that their Global Deal for Nature would cost in the region of US$100 billion per annum (Dinerstein et al. 2019: 14). They do not provide detail on how those figures are calculated, however, other than referring to work by Balmford et al. (2003, 2004) and McCarthy et al. (2012).

As mentioned in Chapter 4, we have reason for scepticism about the ways in which opportunity costs are standardly assessed in the conservation literature and in conservation practice. The use of inappropriate baselines threatens to lock people into poverty, or to take advantage of people's poverty, making conservation 'payments' that are frankly exploitative. For this reason, it is worth looking more closely at the empirical sources that Dinerstein et al. (2019) rely on for their estimate of the costs of protecting half the earth. McCarthy et al. (2012: 947) argue that the costs of conservation in low-income countries are far lower than in high-income countries, because of 'socioeconomic variables'. Balmford et al. (2004: 9697) examine marine protected areas, but the authors are clear, by contrast, that their figures do not include any opportunity costs for local communities. Balmford et al. (2003: 1047), however, do include opportunity costs in their analysis of the costs of increasing protected area coverage, but argue explicitly that opportunity costs correlate closely with 'mean per capita GNP'. It is partly on this basis that they argue that 'field programs typically have far higher benefit-to-cost ratios in less developed parts of the world' (ibid: 1048). In other words, the sources for Dinerstein et al.'s estimate of costs either disregard opportunity costs entirely or assume that opportunity costs are considerably lower in the global South, because of the lower prevailing standards of living there.

Such assumptions are in fact widely shared within the conservation community. One reason conservationists have often supplied for focusing on the tropics is precisely that conservation is so cheap there (the other major reason, of course, is their remarkable concentrations of biodiversity). As Later has put it, policy makers should ask 'in what areas would a given dollar contribute the most toward slowing the current rate of extinction?' (quoted in Youatt 2015: 36). Their conclusion is likely to be that 'environmental capital must be spent where it can get the highest return, in hotspot calculations, and conservation in developed countries is ultimately inefficient by comparison' (Youatt 2015: 36). The worry I express here is that conservation in the tropics may come 'cheap' simply because policy makers are operating an inappropriate baseline for opportunity costs, which places the disadvantage of people in the global South out of normative view (or, worse, because opportunity costs are simply disregarded). In Chapter 4, I argue that conservationists have often been content to exploit locals who possess relatively meagre opportunities, by tying opportunity costs to unjustly straitened opportunities in the contemporary global economy. Unless Half Earth advocates clearly commit to a more demanding baseline, it appears that implementing their proposals could also involve widespread exploitation.

Superficiality

Superficiality may not always be a sin. But in politics, superficiality of analysis can be costly. It can mean our attention is deflected from measures that will actually serve to protect biodiversity, that we are sent down blind alleys, and that we dissipate resources that could have been used to promote conservation or other valuable objectives. Büscher et al. (2017) criticize the Half Earth proposal on the basis that it focuses on the effects of the global biodiversity crisis, but not on its causes. They claim it is naïve to believe that drawing lines around half of the earth would be sufficient if we failed at the same time to tackle the economic processes that actually drive continuing biodiversity loss (see also Napoletano and Clark 2020: 45). Far from doing that, they claim, Wilson appeared to believe that the human economy could simply grow and grow within its own half, and that wise policies would somehow succeed in decoupling growth from further environmental destruction (Büscher and Fletcher 2020: 95).[3] But, they suggest, perhaps a capitalist economy that is sufficiently directed towards environmental (and social) objectives is a chimera.

---

[3] On this topic, Half Earth advocates could usefully be much more forthcoming about their views on biodiversity offsetting. Once large protected areas are established, can they be used to 'offset' the destructive activities of corporations in humanity's half? Chapter 5 suggested there are a number of serious worries about biodiversity offsetting, from the point of view of both global justice and animal rights.

The point is well taken. A key worry here is that what goes on in the 'human' half of the world will inevitably affect the 'natural' half, however strongly the latter might be monitored and patrolled. Climate change is no respecter of fences or juridical boundaries, and neither are many forms of pollution. Each can have a major impact on biodiversity. For Büscher and Fletcher (2020: 96), a failure to accommodate this fact is symptomatic of the way that 'human-oriented activity and processes that affect nature on a global scale are almost completely ignored' in the Half Earth vision. In that sense it may be that Half Earth merely deflects attention from the fact that capitalism is inevitably environmentally destructive. Moreover, we might worry that without serious attention to the growth-oriented dynamics of capitalism itself, all Half Earth would succeed in doing is to *displace* biodiversity destruction from some sites and into others, rather than reducing it overall. Perhaps only an international movement willing and able to constrain the power of international capital could do that (Dempsey and Collard 2017: 37). But Prendergast and Vettese (2021: 18) suggest that Wilson failed to 'marry his scientific insight to a radical economic program able to bring it about', and seemed in fact to possess a rather Promethean faith in capitalism's ability to secure growth without environmental destruction. If his claim was that building fences around Half Earth was *sufficient* to tackle the biodiversity crisis, then there is a danger that such claims will only serve to diminish or deflect the political will needed to tackle the deeper drivers of biodiversity destruction.

Wilson, however, appears as something of an outlier on this issue. For Prendergast and Vettese (2021), a prerequisite for creating a viable Half Earth for nature is the creation of a socialist economy in the human half. Cafaro et al. (2017) and Kopnina et al. (2018) both argue that a secure future for biodiversity will involve at least the reform of capitalism (see also Büscher and Fletcher 2020: 8–9). According to Napoletano and Clark (2020: 38), meanwhile, Half Earth would be best advanced 'as part of a powerful movement capable of unsettling capitalism's instrumental valuation of humans and non-humans alike'.

On this point, many of the most prominent defenders of Half Earth have recently revisited their position, suggesting that they do in fact see Half Earth as only one facet of the conservation coin, and 'transforming human systems' as the other (Crist et al. 2021: 1). Transforming human systems should, they suggest, include a policy of economic degrowth, which might be delivered by way of policies such as 'shortening the workweek, shrinking production of superfluous products . . . lowering the production of animal-derived foods, making commodities that are durable and recyclable', and so on (ibid 2021: 4). They also suggest a focus on human population degrowth, to be achieved exclusively via liberal policies such as expanding the human rights of women and children, providing affordable family planning, guaranteeing secondary education for girls and young women, and bolstering women's rights to buy and inherit property, and to borrow and bank money (ibid 2021: 4–5). To the extent that Half Earth advocates support laudable policies along these lines, they are much less obviously vulnerable to the charge of superficiality.

Political control

Half Earth proponents have been surprisingly silent on the question of who would enjoy political control over Half Earth spaces. Would protected ecoregions be governed by a cadre of conservation scientists? Or would they be governed democratically? If the latter, who would have a stake in their governance—locals, or everyone? Would the Half Earth proposal require qualifications to permanent sovereignty over protected ecoregions, or is it compatible with state resource sovereignty on the current model? The way Half Earth is unpacked would presumably depend partly on whether substantial human populations remain resident in protected ecoregions—a question on which, as we have seen, Half Earth proponents have not always been very upfront. In terms of ideal types, one model of protected area governance would be democratic, with all of those whose significant interests were bound up in the future of a given protected area having a formal stake in decision making about it. Another would be a corporate one, in which decision-making power over protected areas was held by the businesses that sponsored them, or even bought them. In practice, recent years have seen an increase in privatized ownership and control of conservation areas, including ownership across borders (Adams 2020: 792), in part because of the growth of biodiversity offsetting as a policy instrument.

But again, we do not know whether proponents of Half Earth would welcome such arrangements, or whether they believe in a right to participate in decision making about protected areas. Discussions about political control over protected ecoregions have scarcely begun. To the extent that the model continues to influence conservation decision making, critics must continue to press Half Earth advocates to embrace principles of local participation, not only at the level of implementation of protected areas, but also at the level of policy making. Underlying this regrettable silence about political participation is another silence on the topic of knowledge—specifically, what is to count as relevant knowledge about biodiversity. In many cases Half

Earth visions—and this appears especially true of Wilson's writings on the subject—appear to be more or less exclusively grounded in technical data about ecosystem variation, species richness, and so on, but closed to local and indigenous knowledge about biodiversity. In that sense at least, some Half Earth visions threaten to contribute to ongoing epistemic injustices, in which those who actually interact with local biodiversity are not recognized as the possessors of authoritative knowledge about their living environment, with which decision makers will have to reckon (see Tan 2021). More broadly, the ability to generate what will come to be recognized as scientific knowledge of biodiversity is shared highly unequally (Montoya 2022), with some potential knowers occupying privileged positions, even in cases where their efforts turn out to be ultimately parasitic on more local sources of knowledge (see Ghosh 2022).

The GBF has recently suggested that conservation policy makers 'must ensure' that the 'rights, knowledge, including traditional knowledge associated with biodiversity, innovations, worldviews, values and practices of indigenous peoples and local communities are respected, documented, preserved [. . .] including through their full and effective participation in decision-making'. There are good reasons to believe that 'local social-ecological knowledge should be integrated into higher levels of decision making, beyond the community, and into governance at national and even international levels alongside scientific understanding' (Dawson et al. 2021: 11). But it remains to be seen whether indigenous knowledge will be taken into account only once important political decisions have already been made, or whether indigenous views will genuinely make a difference in the wider conservation politics of the future (Ogar et al. 2020). Local participation in conservation should be seen as 'an iterative process requiring time, resources, mutual learning, trust-building and respect for local forms of knowledge and decision-making'—an ongoing challenge rather than a one-off tick-box exercise (Woodhouse et al. 2021: 35). But the need for substantial and sustained local engagement in decision-making has received very little attention from prominent defenders of Half Earth.

Access to green and blue spaces

Some accounts of justice, including my own, hold that the ability to access healthy and diverse environments is an important human interest worthy of protection. That interest might form an element of a wider objective list or hybrid account of well-being, or it might be construed as an important human capability that people should not be deprived of. If so, one important worry about Half Earth is that it might prevent many people from accessing biodiversity. Critics from the global South have argued that the Half Earth proposal—which, like the traditional conservation model, emanates from the global North—enshrines a dualism of nature and the human that is alien to traditional and indigenous communities (Kothari 2021: 161–2). If living in and through biodiversity is important to communities in the South, then—*if* Half Earth embraces a model of exclusion—this will be an important part of the global injustice it creates.

It is also possible, however, that Half Earth would be unfair to people in the global North, and particularly to poor and minority communities living there. If Half Earth led to more intensive economic development within the human half, and if protected ecoregions were relatively few and far between there, this might mean that many people ended up deprived of access to diverse green and blue spaces. This would exacerbate a situation in which many people in rich countries already have inferior access to such spaces, and where access often correlates strongly with class and race. This brings us back to our recurring question about whether Half Earth is to be seen as a stand-alone policy, or one element of a broader package of policies aiming to achieve a wide range of social and environmental objectives. One real danger is that Half Earth might lead to some countries ratcheting down their domestic environmental protection goals. In principle, there would be nothing stopping countries in the North from contributing to the costs associated with running large, protected areas overseas *as well as* safeguarding or even improving access to green and blue spaces at home. But we could also imagine a process whereby governments in rich countries announced that they had discharged their duties towards biodiversity protection by contributing their fair shares towards global conservation efforts, and therefore had less of an obligation to protect biodiversity at home.[4] If access to biodiversity is an important human interest, we should not be satisfied with a 'spectacle of nature', whereby people (whether they are from the North or the South) view biodiversity purely from a distance (see Igoe 2010). To the contrary, taking the interest seriously would speak in favour of efforts to secure the ability to engage in direct physical interaction with biodiversity on a routine basis (Büscher and Fletcher 2020: 171). If Half Earth efforts ended up competing with measures to protect this interest, then this too would represent a serious moral cost.

---

[4] We can also imagine the leaders of Northern countries claiming that their destructive activities were 'offset' by funding protected areas in the South.

Beyond Half Earth

I focus on the Half Earth vision in this chapter both because it is intellectually interesting, and because of its potential political significance. Politically, Half Earth stands as perhaps the grandest proposal for the governance of the world's land and ocean ecosystems ever made. Moreover, it is one with a non-trivial chance of influencing the direction of conservation policy at both global and national levels. While Half Earth is not the only possible model of global conservation governance, the fact that it is both ambitious and influential abundantly justifies closer critical attention. The stakes are high because, as we have seen, Half Earth *could* involve the exclusion of very many people with significant connections to particular places on earth. The history of large-scale, area-based conservation politics has often been elite-driven, poverty-inducing, and exclusionary in nature. It has also frequently embodied a racist disregard for the interests of some people rather than others. Would Half Earth represent a break with that tradition, or a continuation of it? In principle, Half Earth appears to be compatible with either technocratic, elite-led, and exclusionary decision-making processes, or with democratic and open processes that offer everyone whose significant interests are affected an inalienable stake in decision making. To opt for the former when more democratic options are open would also represent a serious political injustice. But Half Earth advocates must be much more transparent about the vision of conservation decision making to which they are committed. They must also reflect more seriously on issues of epistemic injustice, and specifically on the wrongs which can be done to people in their capacity as knowers of biodiversity, and which continue to shape contemporary conservation politics (Tan 2021).

Should our judgement, in the end, be that the Half Earth vision should be endorsed or rejected? This is a difficult question to answer, in part because it should be viewed in *comparative* terms. We must also pay close attention to alternative policies that might advance equivalent goals. If Half Earth is the *only* way of averting the biodiversity crisis, or even, in some sense to be specified, the *best* way, then the proposal might be justified, even if its social, economic, and political effects were significantly negative. By contrast, if alternatives policies are available that are similarly effective, but more compliant with key principles of justice (or more effective, and similarly just), then the pursuit of Half Earth policies would not be defensible. In the end, my view is that policy makers do have access to alternative policies that would be both more effective and more just, and as a result, implementing Half Earth would not be justified. In particular, an alternative package of area-based and non-area-based measures seems to possess considerable promise.

Rethinking area-based conservation

As discussed, over recent decades many post-colonial states have picked up where colonial powers left off in denying customary subsistence and access rights to indigenous and forest-dwelling peoples (Dominguez and Luoma 2020). Elites at both global and national levels have repeatedly advanced their own political ends by presenting locals who are dependent on forests and other ecosystems as the principal threat to biodiversity conservation. Chapter 3 briefly discussed the case of gorilla conservation in the Democratic Republic of the Congo. In that case, the creation of protected areas has been accompanied by the forcible eviction of thousands of indigenous Batwa families, who have also been excluded from local political councils and other decision-making fora (Dominguez and Luoma 2020).

There is encouraging evidence, however, that effective area-based conservation politics do not need to operate in this way. In his cross-national study of forest conservation strategies, Prakash Kashwan has shown that whereas conservation has often meant dispossession and exclusion in India and Tanzania, the same has not been true in Mexico. There, the establishment of protected areas has resulted in 'few, if any displacements or evictions' (Kashwan 2017: 83)—and yet it has been similarly effective. This accords with Brockington and Igoe's well-known survey of conservation-based displacement, which found that evictions are a common feature of conservation in Africa, South and South East Asia, and North America, but not in South and Central America or the Caribbean (Brockington and Igoe 2006: 433). If true, this offers hope that the pathologies often associated with large-scale, area-based conservation policies are contingent and avoidable; further, it suggests that when these pathologies nevertheless come to pass, it is because some policy options better serve elite interests, rather than because they are the only options capable of meeting conservation goals.

A particularly welcome empirical finding over recent years has been that, in many places, 'Indigenous and Community Conserved Areas' are equally successful in preserving biodiversity when contrasted with conventional protected areas. Indigenous peoples officially manage less than 5 per cent of the world's formally protected areas (Dominguez and Luoma 2020). But they have *some* kind of tenure rights over a quarter of the world's land surface (Fa et al. 2020: 135, Tauli-Corpuz et al. 2020: 2). As such, the area under some kind of indigenous control far outstrips the global extent of formally protected areas (Watson et al. 2018b). But what are the prospects for bio-diversity in such areas? One study has found that vertebrate biodiversity in indigenous-managed lands in Australia, Brazil, and Canada matches that in formal protected areas (Schuster et al. 2019). Another has found that indigenous-managed lands in the Peruvian Amazon are *more* effective than state-managed protected areas at preventing deforestation (Schleicher et al. 2017). Other analyses have suggested similar results across Latin America more broadly (Jonas et al. 2014: 113–14). In Asia and Latin America, wildfires have been found to be scarcer in indigenous-managed lands than in official protected areas, and in Brazil rates of deforestation have been found to be lower in the former than the latter (Tauli-Corpuz et al. 2020: 2).

These recent findings are hugely important. To be clear, it is far from certain that the local communities in question will always see themselves as engaged in conservation as an activity. Their priority might simply be to preserve their distinctive ways of living from external threats, as when indigenous people are obliged to defend their life-worlds in the face of extractive industries like logging and mining, even where that defence exposes them to violence, criminalization, and assassination (Scheidel et al. 2020). Indigenous people might even see the idea of 'biodiversity conservation' as a Western concept that meshes poorly with their understanding of the mutually constitutive relationship between themselves and the wider living environment. Whether they see themselves as engaged in preserving biodiversity is perhaps beside the point. What is key is that respecting indigenous control over ecosystems has repeatedly proved amenable to biodiversity preservation, as well as to being conducive to social justice. In fact, formal recognition of, and support for, indigenous-managed areas may actually be one of the most important policy imperatives if policy makers are genuinely committed to meeting goals of biodiversity conservation and global justice simultaneously.[5] An important recent review of the evidence on indigenous-managed land and conventional protected areas has reported some striking findings. Whereas 55.9 per cent of areas controlled by indigenous and local communities reported *both* positive and ecological outcomes, only 15.7 per cent of 'externally controlled' areas (controlled by national or extra-national decision makers) reported both kinds of positive outcome (Dawson et al. 2021: 5). Significantly, the review suggests that the most important factor in

---

[5] One advantage of indigenous-managed areas, as opposed to formal protected areas, has been said to be their relative cheapness: they deliver effective outcomes at a fraction of the cost (Dominguez and Luoma 2020). The arguments of Chapter 4 suggest we treat this claim with some caution, in case it assumes an unjust baseline for calculating any opportunity costs involved.

determining positive conservation outcomes in areas inhabited by indigenous peoples is not the sharing of material benefits (though that surely
also matters, both intrinsically and instrumentally), but 'the recognition of
local social and cultural practices, and the ability of those communities to
influence decision making' (ibid: 6).

As with any argument for greater local control, it is important that
such measures are accompanied by robust protections for minorities within
minorities. It cannot be assumed that indigenous communities are united
around distinct values, or that they are internally cohesive and stable over
time (see Siurua 2006: 77). One important issue is the representation of
women within indigenous decision-making processes (Alvarez and Lovera
2016). Similar points can be made about the participation of people with
diverse sexualities, and religious or ethnic identities. But in principle, these
findings are very encouraging. Both the 2015 Paris Agreement on climate
change and the GBF suggest that countries should expand the tenure rights of
local and indigenous communities. But as of 2020, only 21 out of 131 tropical
countries had formally committed to doing so (Fa et al. 2020: 139). Conservation progress as well as social justice suggests governments must now pay
more than mere lip service to the goal of protecting indigenous land rights
(Kashwan 2017: 216).

Still, even if indigenous-managed areas could cover a quarter of the land
surface of the earth (Fa et al. 2020), this would not get policy makers to
the 50 per cent protection that Half Earth advocates claim is necessary to
stabilize planetary ecosystems. But perhaps what have come to be called
'Other Effective area-based Conservation Measures' (OECMs) hold promise,
too. OECMs are expected to involve a wider set of stakeholders, including
local inhabitants, than conventional protected areas (Weatherley-Singh et al.
2022: 223). Their adoption forms part of a trend in conservation practice
towards 'conserved areas', rather than 'protected' areas—a trend that seems
to run counter to some parts of the Half Earth movement. One review has
found that areas that enhance human well-being by allowing sustainable
resource use tend to have *better* conservation outcomes, compared to areas
that forbid such use (Oldekop et al. 2016). To date, OECMs have largely been
viewed as a possible addition to conventional protected areas, rather than an
alternative to them, and they still represent a very small part of global conservation efforts (Gurney et al. 2021). But it might be that they offer important
advantages compared to conventional protected areas, while being similarly
effective. When combined with indigenous-managed areas, OECMs might
begin to approach the scale of protection sought by Half Earth advocates,
but with far lower social, cultural, and political costs.

Putting global forces in view

To be both effective and just, renewed area-based conservation policies of the type discussed here would need to be embedded in a broader package of policies seeking to address wider human impacts on planetary ecosystems. These impacts derive from unequal and unsustainable patterns of consumption in the global economy, and it is clear that biodiversity cannot be insulated from those impacts by building fences around ecosystems. To date, under the Convention on Biological Diversity framework, targets aiming at expanding protected areas have unfortunately achieved far greater political attention than targets aimed at phasing out harmful subsidies, or reducing over-consumption (Weatherley-Singh et al. 2022: 230). Target 18 of the GBF aims to 'eliminate, phase out or reform incentives, including subsidies, harmful for biodiversity', whereas Target 16 declares that people should be 'encouraged and enabled to make sustainable consumption choices . . . significantly reducing overconsumption'. The project of arresting biodiversity-degrading flows of finance and investment has scarcely begun, however. To date, our understanding of the ecological effects of subsidies is in many cases rudimentary (Dempsey et al. 2020). But it is clear that they often outstrip any positive effects achieved through explicit conservation funding. According to one study, Brazil's US$158 million spend on preventing deforestation, for example, is vastly overshadowed by the US$14 billion it has spent subsidizing activities linked to deforestation (McFarland and Whitley 2015). According to another, subsidies linked to overfishing are more than twice the size of subsidies promoting sustainable fisheries (Sumaila et al. 2019). Reducing these subsidies should be an urgent conservation goal. A crucial intermediate goal will be to weaken the chains of influence between governments and the extractive industries that continue to attract these subsidies, and which often successfully avoid paying for the consequences of their own destructive activities. In this respect campaign finance reform and limits on political lobbying can be important pro-conservation policies (Dempsey and Collard 2017: 38).

Similarly, since indebtedness is often a significant driver of biodiversity destruction in the global South, programmes of debt forgiveness likely have untapped potential to ease destruction (Dempsey et al. 2022). There is evidence that deforestation, for instance, is often driven by indebtedness in the global South (Contestabile 2021), and that levels of debt service correlate positively with extinction risk for mammals and birds (Shandra et al. 2010). The point here is not to advocate 'debt-for-nature' swaps, which have rightly attracted considerable controversy. If the forgiveness of odious debts is already a moral demand, then placing further conditions on forgiveness is inappropriate. Moreover, there are serious worries about whether debt-for-nature swaps have made it still harder for poor states to meet their citizens' basic rights (Hassoun 2012). Nevertheless, forgiving debt—without inappropriate conditions—may well result in the reduction of the fiscal pressures that drive habitat destruction. The connection between debt and biodiversity destruction is too rarely drawn, however, in global conservation politics.

The same point can be made about illegal financial flows, which starve governments in the global South of vital income that could be used for conservation projects (Dempsey et al. 2022: 238). A significant proportion of biodiversity-negative investment appears to be routed through tax havens. On one study, 70 per cent of fishing vessels engaged in illegal, unreported, and unregulated fishing are, or have been, flagged in tax haven jurisdictions, while 68 per cent of the foreign capital invested in keystone corporations associated with deforestation in the Brazilian Amazon has passed through at least one tax haven jurisdiction (Galaz et al. 2018). This suggests that measures to reduce or forgive debt, and to crack down on both illegal financial flows and tax havens, could be key pro-conservation policies.

Of course, Half Earth defenders could argue that they, too, could embrace such measures (even if they have not embraced them in the past). But the point is that the combination of local, participatory conservation measures and radical non-area-based measures appears at least as well equipped to secure effective biodiversity conservation, while *also* protecting vital human interests much better than would Half Earth. If this is true, then dreams of Half Earth should probably be left to one side as an interesting and ambitious thought-experiment that has helped to raise ambition within global conservation politics, but little more than that. Instead, large-scale programmes of indigenous land reform, coupled with concerted efforts to reduce the economic pressures that ultimately drive biodiversity loss, look more likely to serve goals of conservation and global justice simultaneously.

Conclusion

At some point, if our species is to have a future, ordinary citizens and their leaders will have to respond much more earnestly to the challenge of the biodiversity crisis. When they do so, their choice of policies and priorities will have enormous ramifications for global justice. It is possible that a radical conservation politics will help challenge the terrible inequalities of power and privilege that plague our world. But there is also every chance that conservation policies will bolster those inequalities.

There are ample reasons for pessimism about which path is more likely. From the 'White Saviours' of the colonial period—possessed of self-proclaimed moral superiority and astonishingly cavalier about imposing terrible burdens on local populations—through to the armed 'fortresses' of recent decades, often embodying racialized forms of exclusion, the history of conservation practice is a deeply troubling one, with which practitioners and scholars are only slowly getting to grips (Adam 2014). Jessica Dempsey's (2016: 58) claim that a Northern-dominated conservation industry has its gaze locked on the lands of the global South still rings true. Conservation politics is still dominated by governments and NGOs from the North who have finance, technical expertise, and celebrity endorsements at their disposal. People engaged in subsistence activities have often been unfairly portrayed, meanwhile, as the chief enemies of biodiversity, and subjected to coercive and hugely disruptive treatment as a result. That portrayal has served to deflect attention from the environmental impacts of Northern affluence. My focus in this book has generally been forward-looking, but there are important conversations to be had about appropriate collective responses to the unjust history of conservation politics. Apologies, commemorations, and reparations for cases of exclusion and dispossession could all be necessary parts of conservation's process of healing. As the indigenous scholar Cristina Mormorunni suggests, 'Conservation's origin story is finally starting to be told, exposing the fault lines of systemic racism and injustice at its roots, thereby creating the opportunity for equitable and enduring conservation to grow'.[1]

---

[1] https://www.pbs.org/wnet/nature/blog/indigenous-led-conservation/ (Accessed 1 December 2021).

Global Justice and the Biodiversity Crisis. Chris Armstrong, Oxford University Press.
© Chris Armstrong (2024). DOI: 10.1093/9780191888090.003.0008

Other challenges concern the question of whose knowledge and experience is to count within discussions of biodiversity at the national and international levels. Elizabeth Garland (2008: 59) observed that 'There is not a single African who has achieved the kind of global fame from working with African animals that dozens of Western conservationists have attained [. . .] Rarely are they represented as heroic actors on the stage in their own right'. It is shocking that her claims still resonate today. Recent years have witnessed some inspiring visions of truly South-led conservation politics (Ocampo-Ariza et al. 2023) but there is a very long way to go. The academic field of conservation biology continues to be dominated by men from a few countries in the global North (Maas et al. 2021). Researchers from such countries hold a virtual monopoly on science about the deep past of biodiversity (Raja et al. 2022), and wield outsized influence within discussions about its future (Chaudhury and Colla 2021). As Kok-Chor Tan (2021: 7) points out, conservation politics is often characterized by a significant degree of epistemic injustice: all too often 'local knowledge is ignored or disparaged as scientifically invalid', even in cases where it is reproduced, without acknowledgement, in scientific studies.

Failures of inclusion in conservation policy making, and academia, have likely facilitated the imposition of biodiversity policies that are deeply unjust. These policies are frequently unjust in political terms, denying voice to those who ought to be involved in formulating and implementing priorities. And they are frequently unjust in distributive terms too. Policy makers have often followed up on exclusion by offering meagre forms of 'compensation' that do not capture the social, political, and psychological burdens borne by dispossessed locals, nor indeed, the full economic burdens they bear. I argue that conservation programmes have often been frankly exploitative in nature, taking advantage of, and even reinforcing, existing inequalities in access to income and resources. At the same time, I have suggested, the predilection for exclusionary area-based measures has served to deflect attention from less proximate drivers of biodiversity destruction, and specifically from the impact of the consumption habits of the advantaged.

One irony here is that it is now very widely accepted in scholarly and practitioner circles that conservation policies need to become radical, synoptic, and transformative in nature (Wyborn et al. 2020b, Lunquist 2021, Weatherley-Singh et al. 2022). The language of 'transformation' in particular is becoming more or less ubiquitous. It has begun, for example, to percolate into the language of international agreements and the publications of official bodies such as IPBES, which now calls for 'transformative changes across economic, social, political and technological factors' in order to secure biodiversity for the future (IPBES 2019: 14). But it is often difficult to work out what 'transformative' really means in these discussions. The worry is that the term does not really connote any real change in approach, but simply refers to 'business as usual' with a little more ambition grafted on. Alternatively, it *could* indicate policies that engage with the root causes of biodiversity destruction—which means tackling poverty and inequality, as well as wrestling with the intertwined impacts of capitalism and colonialism. It could even refer to fundamental changes in the human interaction with the rest of the living world.

In any case, a discussion of the global justice dimensions of the biodiversity crisis is long overdue. This book attempts to provide a much-needed stimulus to that discussion. It provides an account of biodiversity and why it matters, of the nature and meaning of conservation, and of how we should conceive of, and allocate, conservation burdens. It also stakes out a variety of substantive positions on the global justice dimensions of the biodiversity crisis, revealing the perils and pitfalls of some of the most prominent political responses to date. If we can identify alternatives that promise to be effective in preserving biodiversity without visiting still more injustice on the poor and marginalized, such policies deserve support. I do not pretend, though, that mine are the last words on the topic. My hope is that others will enter the fray and help create the vibrant discussion of global biodiversity justice that we urgently need. I finish by drawing together a number of key conclusions that emerge from the book. Others, though, will no doubt see things that I have not seen, and go on to extend the debate in new directions.

1. Biodiversity loss is accelerating, and collective political responses are failing. The living world has been transformed, and is still being transformed, by processes of capitalism and colonialism. One consequence has been a massive loss of biological diversity, as vibrant, complex ecosystems are replaced by monocrops grown on an industrial scale. Although agriculture-driven habitat loss is the greatest cause of biodiversity loss on land (and destructive fishing practices the biggest problem at sea) the living world is also being pummelled by climate change, pollution, and other challenges. A welcome recent uptick in political ambition—and the increased emphasis among policy makers, scholars, and activists on the need for radical and transformative conservation policies—has not yet arrested the slide towards destruction, some scattered conservation successes notwithstanding.

2. Biodiversity matters hugely to all of our futures. As biodiversity loss continues, it will threaten—in many places it is already threatening—human flourishing and even human survival. It also threatens the interests of non-human animals on a massive scale. The threat it poses to animals might well provide the most important reason for seeking to arrest or reverse bio-diversity loss (while that claim brings us onto contested territory within environmental ethics, it appears to follow readily from several prominent views on the moral status of animals). Biodiversity loss also matters in itself, insofar as biodiversity possesses intrinsic value. For all of these reasons, biodiversity loss demands concerted political attention.

3. *The spectre of a biodiversity crisis has a firm basis in reality.* There is room for debate about whether we have entered, or are about to enter, a Sixth Great Extinction. But there is no doubt that rates of biodiversity loss have accelerated rapidly in the past two centuries—and very rapidly indeed since the Second World War. The language of crisis is entirely relevant when we can observe an empirical process that threatens important interests on a large scale, to which collective political institutions have not yet proven them-selves capable of mounting an effective response. It is not hyperbole to apply the term, then, to processes of biodiversity loss. Despite sustained conserva-tion efforts, biodiversity loss is accelerating. And our knowledge of its likely consequences if unabated is growing continually.

4. *Accepting that the biodiversity crisis demands urgent attention does not require us to believe that biodiversity is all that matters, even when it comes to environmental protection.* Since 1992—the year that witnessed the sign-ing of the Convention on Biological Diversity—collective political responses to environmental destruction have often been organized around the goal of preserving biodiversity. I argue that we do indeed have abundant reasons for tackling the biodiversity crisis. But recognizing this is compatible with accepting that biodiversity is not all that matters, even when it comes to the health of our environment. Other qualities or features of the environment, such as abundance or wildness, can also be appropriate targets of environ-mental policy. Policy makers will sometimes face trade-offs between preserv-ing different environmental goods; any form of 'mission creep' whereby the word biodiversity comes to stand in for the living world as a whole is unhelp-ful. But intelligent policy making ought to be able to handle the idea that environmental politics involves multiple, somewhat competing, objectives.

5. *Political theorists, policy makers, and the wider public have all given insuf-ficient attention to the biodiversity crisis, when compared to other massive environmental problems such as climate change.* If unchecked, the biodiver-sity crisis will threaten the very basis of survival for many millions of people, as well as the other creatures with whom we share the world. It will undercut the conditions for stable economies and stable societies and may exacerbate violent conflict. It will reduce future generations' ability to lead flourishing lives, with the opportunity to engage with a diverse and resilient living world.

None of this is by any means to deny the importance of urgently stepping up our efforts to tackle the climate crisis. But while the biodiversity crisis overlaps with the problem of dangerous climate change, it is also partly distinct from it. Policies aimed at tackling one problem will not automatically solve the other—indeed, there is a danger that measures taken to tackle the climate crisis could have a negative impact on biodiversity (and vice versa). At the very least, this suggests climate policies should be evaluated for their impact on biodiversity (and vice versa). But we must also ask whether institutional responses to the biodiversity crisis—with the Convention on Biological Diversity as their centrepiece—have been in any way adequate. Either way, biodiversity policy making needs to become *genuinely* more ambitious, more synoptic, and more forward looking in its goals, rather than being merely reactive. It also needs to become more systematically joined up with other environmental policy issues. But the need for radical and synoptic policies brings us to our next problem.

6. *Biodiversity conservation poses major, urgent challenges of global justice.* These challenges merit just as much attention as other issues that have come to occupy central roles within the global justice literature, including trade, migration, and climate change. The way policy makers respond to the biodiversity crisis will have profound implications for peoples' opportunities the world over, and this makes it vitally important that principles of justice are explicitly considered when policies are designed and implemented. Poorly designed conservation policies can be unjust in a distributive sense, enabling some to free ride on the sacrifices of others, or allowing policy makers to exploit people living in biodiversity-rich areas. Conservation policies can sustain, and even deepen, injustices of race, class, and gender. They can be politically exclusionary, denying people their rightful input into the design and implementation of those policies, and they can be epistemically unjust, treating some people as deficient in their status as 'knowers' of biodiversity and ecosystems. Showing how conservation policies can be rendered more just is a major challenge, to which scholars from a variety of disciplines will have to contribute.

7. *Conservation burdens are currently shared in a deeply unfair way.* One of the most important facts about contemporary conservation policy—which bears repeating again—is that most of the world's extant biodiversity lies within the global South, while the vast majority of conservation funding is spent in the global North. This misallocation of resources means that some of the world's poorest people are faced with a tragic choice between conservation and escaping from poverty, even in cases where they bear little responsibility for threats to biodiversity, and even in cases where others will

go on to benefit from their conservation activities. Sharing burdens fairly is important in its own right, but it is also instrumentally significant: unfair burden sharing can undermine coalition building, whereas a fair allocation of burdens can sow wider support for conservation. Sharing conservation burdens fairly means paying attention to both our contribution to conservation problems and to our capacity to bear burdens. Moreover, when they refuse to bear their rightful conservation burdens, this will often mean that the affluent are wrongfully free riding on the conservation efforts of poor people in the global South.

8. *Conservation burdens are systematically underestimated but calculating them properly reveals that many conservation schemes exploit the poor.* Burdens need to be shared fairly, but first they need to be defined fairly. Unfortunately, scholars and policy makers often define costs, including opportunity costs, in inappropriate ways. Status quo and willingness-to-accept baselines treat people's existing opportunities as a morally privileged default when it comes to calculating conservation costs. But those opportunities are formed against a backdrop of colonialism, violence, dispossession, and a highly unjust global economic order. Conservation scholars and practitioners must reflect more critically on the kinds of opportunities that people affected by conservation policies *ought* to have and take active measures to ensure that they are not exploiting people's disadvantage. There may be cases where it is simply not feasible to give people greater assistance in dealing with the burdens or upheavals necessitated by conservation. But those cases may be rarer than we think: too often, obtaining conservation on the cheap is a political choice. When they address these issues, policy makers also need to recognize that the losses associated with conservation policies will not always be well captured by financial metrics. But the fact that social, psychological, and cultural impacts cannot be easily captured and compared does not justify ignoring them. Instead, it suggests that formulating conservation policies that are robustly just is an ongoing but vital challenge.

9. *Policy makers should be very cautious about embracing biodiversity offsetting as an alternative to the mitigation or avoidance of biodiversity loss.* Offsetting suffers from a number of problems, some of which are in principle tractable and some of which are not. In principle—though they seem highly reluctant to do so—policy makers could adopt static or positive baselines for biodiversity loss and treat avoidance offsets with a high degree of caution. In principle, they could insist that locals are not excluded from impact or offset sites, and that payments for offsetting more fairly reflect the kinds of opportunities that poor people *ought* to have. But even in a world where policy makers made those choices, offsetting would remain a highly problematic

activity. This is because it appears destined *both* to sever important relationships between people and place, and to condone the infliction of severe
harms on non-human animals. That fact means that it is inappropriate to
treat offsetting as if it were a process in which any morally salient harms are
somehow cancelled by the existence of an offset. Harms remain and must be
taken seriously in any reckoning of the permissibility of offsetting.

10. *The existing governance of protected areas is often deeply problematic.*
The colonial origins of many extant protected areas demand serious reflection. In many cases, post-colonial governments have picked up where colonizers left off in excluding and harassing people—especially tribal peoples—
whose contribution to biodiversity loss is probably comparatively slight, and
who possess few alternatives. Either way, the extension of protected areas
has typically been an elite project, with little input from affected people. I
suggest it is one that often serves to deflect attention from some of the most
important drivers of biodiversity loss. This does not mean that area-based
conservation measures cannot be a part of the future of conservation. But
it is vital to explore alternatives to the traditional fortress model, not least if
other, less-exclusionary models can be just as effective.

11. *There is increasing evidence that the unfinished project of securing the
land rights of indigenous people might be one of the most effective conservation
policies, as well as being socially progressive.* There are abundant reasons of
social and political justice for seeking to promote greater indigenous control
over places of particular significance to their communities. These are rein-
forced by recent findings that indigenous control can be highly conducive to
effective biodiversity protection, matching—even outperforming—the successes of conventional protected areas. Elites often find reasons for scepticism
about granting greater autonomy to indigenous communities. In India and
Tanzania, indigenous communities have often been presented as the enemies
of conservation. But those claims probably have more to do with the desire
to maintain elite control than they do with empirical reality. As evidence
about the ecosystem vitality of many indigenous-controlled areas mounts,
the argument from both conservation and global justice appears stronger
and stronger. As well as promoting indigenous land claims, policy makers
should promote their full participation at all levels of conservation politics,
from design through to implementation and beyond.

12. *It is hard to see how many Half Earth visions can be justified.* The
Half Earth project is increasingly influential and has some chance of shifting the overall direction of global conservation policy. Nevertheless, I have
demonstrated that defenders of Half Earth have major questions to answer
when it comes to the possible global justice impacts of their proposals. These

questions revolve around, for instance, exclusion, deprivation, and the superficiality of some Half Earth proposals. They also invoke issues of political control and epistemic justice. More radical defenders of Half Earth have presented some answers to these worries, but those answers have not been entirely satisfying. Perhaps the argument most damaging to their cause is that rival projects, such as entrenching indigenous land rights, promise to be at least as effective at protecting biodiversity over large areas of land while having much more favourable implications for social, political, and distributive justice. At the same time, conservation policy makers ought to give much more attention to effective non-area-based conservation policies. If such measures, combined with greater protection for indigenous land claims, can be effective, this appears to have dramatic implications for the permissibility of the Half Earth project.

13. *There has been an excessive focus on the geographically proximate causes of biodiversity loss.* In many cases, biodiversity loss is driven by action at a distance—as in the case of climate change, where one person's emissions could contribute to ecosystem damage anywhere in the world. In other cases, the causes are more proximate: destructive fishing practices and deforestation are carried out *in situ* by individuals and corporations. But neither land-use change nor the over-exploitation of local ecosystems occur in a vacuum. To the contrary: they are often inextricably linked to resource flows and opportunity structures in the wider global economy, to indebtedness, to tax competition, to the scale of illicit financial flows, and to many other exogenous pressures on local decision makers. Policy makers have sometimes acted as if the creation of protected areas is the only intelligible solution to biodiversity destruction. But to focus exclusively on proximate causes—and to adopt only proximate responses—is myopic and unhelpful. Further, it is also frequently unjust to boot. Relying on coercive and exclusionary policies that leave burdens on the shoulders of those who can least afford to bear them cannot be justified, if other effective responses are available that do not have such regressive consequences for the poor and marginalized.

14. *Conservation policy makers would do well to focus much more on global and systemic pressures towards biodiversity destruction.* Pressuring local people to desist from destroying biodiversity, in cases where they have few options but to do so, is probably futile (as well as unfair, if it means they are expected to bear the burdens of conservation). Poor people often care just as much about biodiversity as wealthy people do, albeit they may not recognize simplistic (and largely Western) divisions between humanity and nature, society and wilderness. In many cases it is likely to be more effective, as well as fairer, to focus on actors and incentives higher up 'value

chains' in the global economy, far beyond the sites where primary extraction occurs. Policy makers must recognize that reducing harmful subsidies, rolling back investment in fossil fuels and meat production, and tackling illicit financial flows, are viable conservation policies in their own right—and policies, moreover, that may well be both fairer and more effective than area-based measures. They should also aim to reduce the political power of industries harmful to biodiversity—such as industrial agriculture, and fossil fuels—by pursuing limits on political donations and the power of lobbying. Ultimately, though, the most effective pro-conservation policies might be policies aimed at reducing global inequality. Through processes of environmental load displacement, the advantaged have so far proven quite adept at protecting themselves from the environmental consequences of their consumption. But evidence suggests that inequality is a key driver of biodiversity loss, at both national and global levels. Measures aimed at reducing the environmental footprints of the advantaged therefore have to be a key part of any transformative conservation politics.

15. *Preserving biodiversity in the global South might be considered economically 'efficient', but we need to protect biodiversity everywhere.* In reality, policy makers' judgements about efficiency are unduly influenced by the adoption of highly dubious baselines for calculating conservation costs. The reason conservation in the global South looks cheap is often simply that policy makers assume standards of living in the South are, and can remain, very modest. This assumption needs tackling. But even if conservation in the global South *was* more efficient, this does not mean that we should be content with a scenario where biodiversity is protected there and allowed to dwindle everywhere else. People everywhere have an interest in being able to access and interact with biodiverse ecosystems (of which they are, of course, a part). Major inequalities of access to biodiversity in the global North are also a problem of justice, which must not be compounded by conservation policies that focus excessively on the bottom line. In many cases, policy makers might face difficult trade-offs—where, given the same budget, they can protect more biodiversity further away from people, or less biodiversity closer to them. The first step is to recognize that these *are* genuine trade-offs, and that conservation policy should not focus on efficiency alone. If access to biodiversity is important for human flourishing, then patterns of access also matter.

Bibliography

Abegão, João L. R. 2019. 'Where the Wild Things Were Is Where Humans Are Now: An Overview', *Human Ecology* 47.5: 669–79.

Adam, Rachelle. 2014. *Elephant Treaties: The Colonial Legacy of the Biodiversity Crisis* (London: University Press of New England).

Adams, Vanessa, Robert Pressey, and Robin Naidoo. 2010. 'Opportunity Costs: Who Really Pays for Conservation?', *Conservation Biology* 143: 439–48.

Adams, William et al. 2004. 'Biodiversity Conservation and the Eradication of Poverty', *Science* 306.5699: 1146–9.

Adams, William. 2020. 'Geographies of Conservation III: Nature's Spaces', *Progress in Human Geography* 44.4: 789–801.

Agarwal, Anil and Sunita Narain. 1991. *Global Warming in an Unequal World: A Case of Environmental Colonialism* (New Delhi: Centre for Science and Environment).

Allan, James et al. 2019. 'Hotspots of Human Impact on Threatened Terrestrial Vertebrates', *PLoS Biology*, 17(3), e3000158.

Allan, James et al. 2022. 'The Minimum Land Area Requiring Conservation Attention to Safeguard Biodiversity', *Science* 376.6597: 1094–101.

Alvarez, Isis and Simone Lovera. 2016. 'New Times for Women and Gender Issues in Biodiversity Conservation and Climate Justice', *Development* 59: 263–5.

Anderson, Christopher B. et al. (2022). 'Conceptualizing the Diverse Values of Nature and Their Contributions to People', in P. Balvanera et al. (eds) *Methodological Assessment Report on the Diverse Values and Valuation of Nature* (Bonn: IPBES Secretariat), 36–121.

Apostolopoulou, Evangelia and William Adams. 2017. 'Biodiversity Offsetting and Conservation: Reframing Nature to Save It', *Oryx* 51.1: 23–31.

Arlidge, William et al. 2018. 'A Global Mitigation Hierarchy for Nature Conservation', *BioScience* 68.5: 336–47.

Armstrong, Chris. 2012. *Global Distributive Justice: An Introduction* (Cambridge: Cambridge University Press).

Armstrong, Chris. 2014. 'Justice and Attachment to Natural Resources', *Journal of Political Philosophy* 22.1: 48–65.

Armstrong, Chris. 2015. 'Against Permanent Sovereignty over Natural Resources', *Politics, Philosophy & Economics* 14.2: 129–51.

Armstrong, Chris. 2016. 'Fairness, Free Riding and Rainforest Protection', *Political Theory* 44: 106–30.

Armstrong, Chris. 2017a. *Justice and Natural Resources: An Egalitarian Theory* (Oxford: Oxford University Press).

Armstrong, Chris. 2017b. 'Climate Change and Justice', *Oxford Research Encyclopedia of Politics* [online], 20 November. https://doi.org/10.1093/acrefore/9780190228637.013.231

Armstrong, Chris. 2019a. *Why Global Justice Matters: Moral Progress in a Divided World* (Cambridge: Polity).

Armstrong, Chris. 2019b. 'Sharing Conservation Burdens Fairly', *Conservation Biology* 33.3: 554–60.

Armstrong, Chris. 2020. 'Decarbonisation and World Poverty: A Just Transition for Fossil Fuel Exporting Countries?', *Political Studies* 68.3: 671–88.

Armstrong, Chris. 2021. 'Natural Resources, Sustainability, and Intergenerational Ethics', in Stephen Gardiner (ed) *Oxford Handbook of Intergenerational Ethics* (Oxford: Oxford University Press), ch. 19.

Armstrong, Chris. 2022. *A Blue New Deal: Why We Need a New Politics for the Ocean* (New Haven: Yale University Press).

Armstrong, Chris. 2023. 'The United Nations Convention on the Law of the Sea, Global Justice, and the Environment', to be published in *Global Constitutionalism* [preprint].

Armstrong, Chris and Duncan McLaren. 2022. 'Which Net Zero? Climate Justice and Net Zero Emissions', *Ethics & International Affairs* 36.4: 505–26.

Austin, Kelly. 2021. 'Degradation and Disease: Ecologically Unequal Exchanges Cultivate Emerging Pandemics', *World Development* 137: 105163.

Baard, Patrik. 2021. 'Biocentric Individualism and Biodiversity Conservation: An Argument from Parsimony', *Environmental Values* 30.1: 93–110.

Balmford, Andrew and Tony Whitten. 2003. 'Who Should Pay for Tropical Conservation, and How Could the Costs Be Met?' *Oryx* 37.2: 238–50.

Balmford, Andrew et al. 2003. 'Global Variation in Terrestrial Conservation Costs, Conservation Benefits, and Unmet Conservation Needs', *Proceedings of the National Academy of Sciences of the United States of America* 100.3: 1046–50.

Balmford, Andrew et al. 2004. 'The Worldwide Costs of Marine Protected Areas', *Proceedings of the National Academy of Sciences of the United States of America* 101.26: 9694–7.

Bar-On, Yinon, Rob Phillips, and Ron Milo. 2018. 'The Biomass Distribution on Earth', *Proceedings of the National Academy of Sciences of the United States of America* 115.25: 6506–11.

Barrett, Christopher, Alexander Travis, and Partha Dasgupta. 2011. 'On Biodiversity Conservation and Poverty Traps', *Proceedings of the National Academy of Sciences of the United States of America* 108.34: 13907–12.

Barry, Christian and Garret Cullity. 2022. 'Offsetting and Risk Imposition', *Ethics* 132.2: 352–81.

Barry, Christian and Gerhard Øverland. 2015. 'Individual Responsibility for Carbon Emissions: Is There Anything Wrong with Overdetermining Harm?' in Jeremy Moss (ed) *Climate Change and Justice* (Cambridge: Cambridge University Press), 165–83.

Barry, Christian and Gerhard Øverland. 2016. *Responding to Global Poverty* (Cambridge: Cambridge University Press).

Batavia, Chelsea and Michael Paul Nelson. 2017. 'Heroes or Thieves? The Ethical Grounds for Lingering Concerns about New Conservation', *Journal of Environmental Studies and Sciences* 7.3: 394–402.

Bidaud, Cecile et al. 2017. 'The Sweet and the Bitter: Intertwined Positive and Negative Social Impacts of a Biodiversity Offset', *Conservation and Society* 15.1: 1–13.

Bidaud, Cécile, Kate Schreckenberg, and Julia Jones. 2018. 'The Local Costs of Biodiversity Offsets: Comparing Standards, Policy and Practice', *Land Use Policy* 77: 43–50.

Billé, Raphaël, Gilles Kleitz, and Gregory Mikkelson. 2013. 'Economic Equality as a Condition for Biodiversity Conservation', in Remi Genevey, Tancrede Voituriez, and Sanjivi Sundar (eds) *Reducing Inequalities: A Sustainable Development Challenge* (Delhi: TERI Press), 101–18.

Bolam, Friederike et al. 2021. 'How Many Bird and Mammal Extinctions Has Recent Conservation Action Prevented?' *Conservation Letters* 14.1: e12762.

Bremer, Leah and Kathleen Farley. 2010. 'Does Plantation Forestry Restore Biodiversity or Create Green Deserts? A Synthesis of the Effects of Land-use Transitions on Plant Species Richness', *Biodiversity and Conservation* 19.14: 3893–915.

Britton, Easkey et al. 2020. 'Blue Care: A Systematic Review of Blue Space Interventions for Health and Wellbeing', *Health Promotion International* 35.1: 50–69.

Broadhead, Stearns and Adriana Placani. 2021. 'The Morality of Offsets for Luxury Emissions', *World Futures* 77.6: 405–17.

Brockington, Dan. 2002. *Fortress Conservation: The Preservation of the Mkomazi Game Reserve, Tanzania* (Bloomington: Indiana University Press).

Brockington, Dan. 2004. 'Community Conservation, Inequality, and Injustice: Myths of Power in Protected Area Management', *Conservation and Society* 2.2: 411–32.

Brockington, Dan and James Igoe. 2006. 'Eviction for Conservation: A Global Overview', *Conservation and Society* 4.3: 424–70.

Broome, John. 2012. *Climate Matters: Ethics in a Warming World* (New York: W.W. Norton).

Broome, John. 2019. 'Against Denialism', *The Monist* 102.1: 110–29.

Bull, Joseph and Niels Strange. 2018. 'The Global Extent of Biodiversity Offset Implementation under No Net Loss Policies', *Nature Sustainability* 1.12: 790–8.

Buller, Adrienne. 2022. *The Value of a Whale: On the Illusions of Green Capitalism* (Manchester: Manchester University Press).

Burch-Brown, Joanna and Alfred Archer. 2017. 'In Defence of Biodiversity', *Biology & Philosophy* 32.6: 969–97.

Büscher, Bram et al. 2017. 'Half-Earth or Whole Earth? Radical Ideas for Conservation, and Their Implications', *Oryx* 51.3: 407–10.

Büscher, Bram and Robert Fletcher. 2016. 'Why EO Wilson is Wrong about How to Save the Earth', *Aeon* [online] 1 March. https://aeon.co/ideas/why-eo-wilson-is-wrong-about-how-to-save-the-earth

Büscher, Bram and Robert Fletcher. 2020. *The Conservation Revolution: Radical Ideas for Saving Nature Beyond the Anthropocene* (London: Verso).

Bush, Glenn, et al. 2013. 'Measuring the Local Costs of Conservation: A Provision Point Mechanism for Eliciting Willingness to Accept Compensation', *Land Economics* 89.3: 490–513.

Cafaro, Philip et al. 2017. 'If We Want a Whole Earth, Nature Needs Half: A Response to Büscher et al.', *Oryx* 51.3: 400.

Cafaro, Philip et al. 2022. 'Overpopulation Is a Major Cause of Biodiversity Loss and Smaller Human Populations are Necessary to Preserve What Is Left', *Biological Conservation* 272: 109646.

Callicott, J. Baird and Michael Nelson (eds). 1998. *The Great New Wilderness Debate* (Athens, GA: University of Georgia Press).

Calvet, Coralie et al. 2015. 'The Biodiversity Offsetting Dilemma: Between Economic Rationales and Ecological Dynamics', *Sustainability* 7.6: 7357–78.

Caney, Simon. 2005. *Justice Beyond Borders* (Oxford: Oxford University Press).

Caney, Simon. 2009. 'Climate Change, Human Rights and Moral Thresholds', in Charles Beitz and Robert Goodin (eds) *Global Basic Rights* (Oxford: Oxford University Press), 227–47.

Caney, Simon. 2010. 'Climate Change and the Duties of the Advantaged', *Critical Review of International Social and Political Philosophy* 13.1: 203–28.

Caney, Simon. 2012a. 'Addressing Poverty and Climate Change: The Varieties of Social Engagement', *Ethics & International Affairs* 26.2: 191–216.

Caney, Simon. 2012b. 'Just Emissions', *Philosophy & Public Affairs* 40.4: 255–300.

Caney, Simon. 2014. 'Two Kinds of Climate Justice: Avoiding Harm and Sharing Burdens', *Journal of Political Philosophy* 22.2: 125–49.

Caney, Simon. 2020. 'Human Rights, Population, and Climate Change', in Dapo Akande et al. (eds) *Human Rights and 21st Century Challenges: Poverty, Conflict, and the Environment* (Oxford: Oxford University Press), 348–69.

Caney, Simon and Cameron Hepburn. 2011. 'Carbon Trading: Unethical, Unjust and Ineffective?' *Royal Institute of Philosophy Supplement* 69: 201–34.

Cardinale, Bradley et al. 2012. 'Biodiversity Loss and Its Impact on Humanity', *Nature* 486.7401: 59–67.

Chaudhury, Aadita and Sheila Colla. 2021. 'Next Steps in Dismantling Discrimination: Lessons from Ecology and Conservation Science', *Conservation Letters* 14.2: e12774.

Christensen, Villy et al. 2014. 'A Century of Fish Biomass Decline in the Ocean', *Marine Ecology Progress Series* 512: 155–66.

Clay, Jason. 2011. 'Freeze the Footprint of Food', *Nature* 475.7356: 287–89.

Clement, Matthieu and Andre Meunie. 2010. 'Is Inequality Harmful for the Environment? An Empirical Analysis Applied to Developing and Transition Countries', *Review of Social Economy* 68.4: 413–45.

Cochrane, Alasdair. 2018. *Sentientist Politics: A Theory of Global Inter-Species Justice* (Oxford: Oxford University Press).

Cohen, Gerald A. 2013. *Finding Oneself in the Other*, Michael Otsuka (ed) (Princeton: Princeton University Press).

Collins, Stephanie. 2019. *Group Duties: The Existence and Their Implications for Individuals* (Oxford: Oxford University Press).

Contestabile, Monica. 2021. 'Joined-up Action for Biodiversity', *Nature Sustainability* 4.8: 660–1.

Corlett, Richard. 2020. 'Safeguarding our Future by Protecting Biodiversity', *Plant Diversity* 42: 221–8.

Cripps, Elizabeth. 2022. *What Climate Justice Means and Why We Should Care* (London: Bloomsbury).

Crist, Eileen et al. 2021. 'Protecting Half the Planet and Transforming Human Systems are Complementary Goals', *Frontiers in Conservation Science* 2: 761292.

Cronon, William. 1996. 'The Trouble with Wilderness: Or, Getting Back to the Wrong Nature', *Environmental History* 1.1: 7–28.

Crutzen, Paul. 2006. 'The "Anthropocene"', in Eckart Ehlers and Thomas Krafft (eds) *Earth System Science in the Anthropocene*. (Berlin: Springer), 13–18.

Curran, Michael et al. 2014. 'Is There Any Empirical Support for Biodiversity Offset Policy?' *Ecological Applications* 24.4: 617–32.

Cushing, Lara et al. 2015. 'The Haves, the Have-nots, and the Health of Everyone: The Relationship between Social Inequality and Environmental Quality', *Annual Review of Public Health* 36: 193–209.

Dai, Dajun. 2011. 'Racial/ethnic and Socioeconomic Disparities in Urban Green Space Accessibility: Where to Intervene?' *Landscape and Urban Planning* 102.4: 234–44.

Damiens, Florence, Libby Porter, and Ascelin Gordon. 2021. 'The Politics of Biodiversity Offsetting across Time and Institutional Scales', *Nature Sustainability* 4.2: 170–9.

Dasgupta, Partha. 2021. *The Economics of Biodiversity: The Dasgupta Review* (London: HM Treasury).

Dawson, Neil et al. 2021. 'The Role of Indigenous Peoples and Local Communities in Effective and Equitable Conservation', *Ecology and Society* 26.3: 1–18.

DEFRA, 2019. *Net Gain: Summary of Responses and Government Response* (London: Department for Environment, Farming and Rural Affairs).

Dempsey, Jessica. 2016. *Enterprising Nature: Economics, Markets, and Finance in Global Biodiversity Politics* (Chichester: John Wiley & Sons).

Dempsey, Jessica et al. 2020. 'Subsidizing Extinction?' *Conservation Letters* 13: e12705.

Dempsey, Jessica et al. 2022. 'Biodiversity Targets Will Not Be Met without Debt and Tax Justice', *Nature Ecology & Evolution* 6: 237–39.

Dempsey, Jessica and Rosemary-Claire Collard. 2017. 'If Biodiversity Offsets Are a Dead End for Conservation, what Is the Live Wire? A Response to Apostolopoulou & Adams', *Oryx* 51.1: 35–9.

Deplazes-Zemp, Anna. 2019. 'A Global Biodiversity Fund to Implement Distributive Justice for Genetic Resources', *Developing World Bioethics* 19.4: 235–44.

Deplazes-Zemp, Anna. 2023. 'Beyond Intrinsic and Instrumental: Third-category Value in Environmental Ethics and Environmental Policy', *Ethics, Policy & Environment* [online] https://doi.org/10.1080/21550085.2023.2166341

De Silva, Sunila and Krithika Srinivasan. 2019. 'Revisiting Social Natures: People–elephant Conflict and Coexistence in Sri Lanka', *Geoforum* 102: 182–90.

Devault, Damien et al. 2021. 'The Silent Spring of Sargassum', *Environmental Science and Pollution Research* 28: 15580–3.

De Vos, Jurriaan et al. 2015. 'Estimating the Normal Background Rate of Species Extinction', *Conservation Biology* 29.2: 452–62.

Díaz, Sandra et al. 2006. 'Biodiversity Loss Threatens Human Well-Being', *PLoS Biology* 4.8: 1300–5.

Dinda, Soumyananda. 2004. 'Environmental Kuznets Curve Hypothesis: A Survey', *Ecological Economics* 49.4: 431–55.

Dinerstein, Eric et al. 2017. 'An Ecoregion-based Approach to Protecting Half the Terrestrial Realm', *BioScience* 67.6: 534–45.

Dinerstein, Eric, et al. 2019. 'A Global Deal for Nature: Guiding Principles, Milestones, and Targets', *Science Advances* 5.4: eaaw2869.

Dinerstein, Eric et al. 2020. 'A "Global Safety Net" To Reverse Biodiversity Loss and Stabilize Earth's Climate', *Science Advances* 6.36: eabb2824.

Doak, Daniel et al. 2015. 'What Is the Future of Conservation?' in George Wuerthner, Eileen Crist, and Tom Butler (eds) *Protecting the Wild* (Washington DC: Island Press), 27–35.

Dominguez, Lara and Colin Luoma. 2020. 'Decolonising Conservation Policy: How Colonial Land and Conservation Ideologies Persist and Perpetuate Indigenous Injustices at the Expense of the Environment', *Land* 9.3: 65.

Donaldson, Sue and Will Kymlicka. 2011. *Zoopolis: A Political Theory of Animal Rights* (Oxford: Oxford University Press).

Dowie, Mark. 2009. *Conservation Refugees: The Hundred-year Conflict between Global Conservation and Native Peoples* (Cambridge, MA: MIT Press).

Dowle, Eddy. 2013. 'Molecular Evolution and the Latitudinal Biodiversity Gradient', *Heredity* 110: 501–10.

Dryzek, John and Jonathan Pickering. 2019. *The Politics of the Anthropocene* (Oxford: Oxford University Press).

Duffy, Rosaleen. 2014. 'Waging a War to Save Biodiversity: The Rise of Militarized Conservation', *International Affairs* 90.4: 819–34.

Duffy, Rosaleen et al. 2019. 'Why We Must Question the Militarisation of Conservation', *Biological Conservation* 232: 66–73.

Ebel, Roland et al. 2021. 'How Biodiversity Loss Affects Society', in Harvey James (ed) *Handbook on the Human Impact of Agriculture* (Cheltenham: Edward Elgar Publishing), 352–76.

Eckersley, Robyn. 1992. *Environmentalism and Political Theory: Toward an Ecocentric Approach* (New York: Suny Press).

Ekins, Paul. 2003. 'Identifying Critical Natural Capital: Conclusions about Critical Natural Capital', *Ecological Economics* 44.2–3: 277–92.

Elliot, Robert. 1992. 'Intrinsic Value, Environmental Obligation, and Naturalness', *The Monist* 75.2: 138–60.

Ellis, Erle and Zia Mehrabi. 2019. 'Half Earth: Promises, Pitfalls, and Prospects of Dedicating Half of Earth's Land to Conservation', *Current Opinion in Environmental Sustainability* 28: 22–30.

Fa, Julia et al. 2020. 'Importance of Indigenous Peoples' Lands for the Conservation of Intact Forest Landscapes', *Frontiers in Ecology and the Environment* 18.3: 135–40.

Fabre, Cécile. 2021. 'Territorial Sovereignty and Humankind's Common Heritage', *Journal of Social Philosophy* 52.1: 17–23.

Fabre, Cécile. 2022. 'To Snatch Something from Death': Value, Justice, and Humankind's Common Cultural Heritage', *Tanner Lectures in Human Values*, 10 May [online].

Faith, Daniel. 2017. 'A General Model for Biodiversity and Its Value', in Justin Garson, Anya Plutynski, and Sahotra Sarkar (eds) *Routledge Handbook of Philosophy of Biodiversity* (London: Routledge), 69–85.

Farnham, Timothy. 2017. 'A Confluence of Values: Historical Roots of Concern for Biological Diversity', in Justin Garson, Anya Plutynski, and Sahotra Sarkar (eds) *Routledge Handbook of Philosophy of Biodiversity* (London: Routledge), 11–25.

Feinberg, Joel. 2017. 'The Rights of Animals and Unborn Generations', in Russ Shafer-Landau (ed) *Ethical Theory: An Anthology* (Chichester: Wiley-Blackwell), 372–80.

Ferns, George. 2022. 'Businesspeople Must Reconnect with Nature to Save the Planet', *MIT Sloan Management Review*, 21 April [online].

Fischer, Anke and Juliette Young. 2007. 'Understanding Mental Constructs of Biodiversity: Implications for Biodiversity Management and Conservation', *Biological Conservation* 136.2: 271–82.

Fischer, Joern et al. 2017. 'Reframing the Food-biodiversity Challenge', *Trends in Ecology and Evolution* 32.5: 335–45.

Fisher, Brendan et al. 2011. 'Implementation and Opportunity Costs of Reducing Deforestation and Forest Degradation in Tanzania', *Nature Climate Change* 1.3: 161–4.

Fraser, Nancy. 2021. 'Climates of Capital: For a Trans-environmental Eco-Socialism', *New Left Review* 127: 94–127.

Galaz, Victor et al. 2018. 'Tax Havens and Global Environmental Degradation', *Nature Ecology & Evolution* 2: 1352–7.

García-Portela, Laura. 2023. 'Backward-Looking Principles of Climate Justice: The Unjustified Move from the Polluter Pays Principle to the Beneficiary Pays Principle', *Res Publica* 29: 367–84.

Gardiner, Stephen. 2011. *A Perfect Moral Storm: The Ethical Tragedy of Climate Change* (Oxford: Oxford University Press).

Garland, Elizabeth. 2008. 'The Elephant in the Room: Confronting the Colonial Character of Wildlife Conservation in Africa', *African Studies Review* 51.3: 51–74.

Gheaus, Anca. 2019. 'More Co-parents, Fewer Children: Multiparenting and Sustainable Population', *Essays in Philosophy* 20.1: 3–23.

Ghosh, Amitav. 2022. *The Nutmeg's Curse: Parables for a Planet in Crisis* (London: John Murray).

Gilabert, Pablo. 2017. 'Justice and Feasibility: A Dynamic Approach', in Kevin Vallier and Michael Weber (eds) *Political Utopias: Contemporary Debates* (Oxford: Oxford University Press), 95–126.

Global Forest Coalition. 2022. *Net Gain is a Lose–lose for Rights, Gender Justice and Social Equity in Biodiversity Policy* (Asunción, Paraguay: Global Forest Coalition).

Gordon, Ascelin et al. 2015. 'Perverse Incentives Risk Undermining Biodiversity Offset Policies', *Journal of Applied Ecology* 52.2: 532–37.

Gosseries, Axel. 2015. 'What's Wrong with Trading Emission Rights', in Jeremy Moss (ed) *Climate Change and Justice* (Cambridge: Cambridge University Press), 89–106.

Graeber, David and David Wengrow. 2021. *The Dawn of Everything: A New History of Humanity* (London: Penguin).

Granjon, A.-C. et al. 2020. 'Estimating Abundance and Growth Rates in a Wild Mountain Gorilla Population', *Animal Conservation* 23.4: 455–65.

Green, Fergus. 2021. 'Ecological Limits: Science, Justice, Policy, and the Good Life', *Philosophy Compass* 16.6: e12740.

Green, Jonathan et al. 2018. 'Local Costs of Conservation Exceed those Borne by the Global Majority', *Global Ecology and Conservation* 14: e00385.

Guha, Ramachandra. 1989. 'Radical American Environmentalism and Wilderness Preservation: A Third World Critique', *Environmental Ethics* 11.1: 71–83.

Gurney, Georgina et al. 2021. 'Biodiversity Needs Every Tool in the Box: Use OECMs', *Nature* 595: 646–49.

Hale, Benjamin. 2016. 'Rights, Rules, and Respect for Nature', in Stephen M. Gardiner and Allen Thompson (eds) *The Oxford Handbook of Environmental Ethics* (Oxford: Oxford University Press), 211–22.

Hall, Richard, E. J. Milner-Gulland, and F. Courchamp. 2008. 'Endangering the Endangered: The Effects of Perceived Rarity on Species Exploitation', *Conservation Letters* 1.2: 75–81.

Hamann, Maike et al. 2018. 'Inequality and the Biosphere', *Annual Review of Environment and Resources* 43: 61–83.

Haraway, Donna. 2015. 'Anthropocene, Capitalocene, Plantationocene, Chthulucene: Making Kin', *Environmental Humanities* 6.1: 159–65.

Hassoun, Nicole. 2012. 'The Problem of Debt-for-Nature Swaps from a Human Rights Perspective', *Journal of Applied Philosophy* 29.4: 359–77.

Hassoun, Nicole and David Wong. 2015. 'Conserving Nature, Preserving Identity', *Journal of Chinese Philosophy* 42.1–2: 176–96.

Hassoun, Nicole et al. 2020. 'Multidimensional Poverty Measurement: The Value of Life and the Challenge to Value Aggregation', in Valentin Beck, Henning Hahn, and Robert Lepenies (eds) *Dimensions of Poverty: Measurement, Epistemic Injustices, Activism* (Dordrecht: Springer), 321–37.

Hillebrand, Helmut. 2004. 'On the Generality of the Latitudinal Diversity Gradient', *The American Naturalist* 163.2: 192–211.

Holland, Tim, Garry Peterson, and Andrew Gonzalez. 2009. 'A Cross-national Analysis of How Economic Inequality Predicts Biodiversity Loss', *PLoS One* 2: e444.

Holmes, George, Chris Sandbrook, and Janet Fisher. 2017. 'Understanding Conservationists' Perspectives on the New-Conservation Debate', *Conservation Biology* 31.2: 353–63.

Holtug, Nils. 2007. 'Equality for Animals', in Jesper Ryberg, Thomas Petersen, and Clark Wolf (eds) *New Waves in Applied Ethics* (Basingstoke: Palgrave Macmillan), 1–24.

Hornborg, Alf. 2006. 'Footprints in the Cotton Fields: The Industrial Revolution as Time-Space Appropriation and Environmental Load Displacement', *Ecological Economics* 59.1: 74–81.

Horta, Oscar. 2016. 'Egalitarianism and Animals', *Between the Species* 19.1: 109–45.

Hudson, Lawrence et al. 2017. 'The Database of the PREDICTS (Projecting Responses of Ecological Diversity in Changing Terrestrial Systems) Project', *Ecology and Evolution* 7.1: 145–88.

Hulme, David and Marshall Murphree. 1999. 'Communities, Wildlife and the "New Conservation" in Africa', *Journal of International Development* 11: 277–85.

Human Rights Council. 2017. *Report of the Special Rapporteur on the Issue of Human Rights Obligations Relating to the Enjoyment of a Safe, Clean, Healthy and Sustainable Environment.* A/HRC/34/49 (19 January 2017).

Hyams, Keith and Tina Fawcett. 2013. 'The Ethics of Carbon Offsetting', *WIRES Climate Change* 4.2: 91–98.

Igoe, Jim. 2010. 'The Spectacle of Nature in the Global Economy of Appearances: Anthropological Engagements with the Spectacular Mediations of Transnational Conservation', *Critique of Anthropology* 30.4: 375–97.

IPBES. 2019. *Summary for Policymakers of the Global Assessment Report on Biodiversity and Ecosystem Services of the Intergovernmental Science-policy Platform on Biodiversity and Ecosystem Services* (Bonn, Germany: IPBES Secretariat).

Ives, Christopher and Sarah Bekessy. 2015. 'The Ethics of Offsetting Nature', *Frontiers in Ecology and the Environment* 13.10: 568–73.

Jonas, Harry et al. 2014. 'New Steps of Change: Looking beyond Protected Areas to Consider Other Effective Area-based Conservation Measures', *Parks* 20.2: 111–28.

Kagan, Shelly. 2018. 'For Hierarchy in Animal Ethics', *Journal of Practical Ethics* 6.1: 1–18.

Kallhoff, Angela. 2021. 'The Intergenerational Value of Natural Heritage', in Stephen Gardiner (ed) *Oxford Handbook of Intergenerational Ethics* (Oxford: Oxford University Press), https://doi.org/10.1093/oxfordhb/9780190881931.013.25

Kant, Immanuel. 2017. *The Metaphysics of Morals* (Cambridge: Cambridge University Press).

Kareiva, Peter and Michelle Marvier. 2007. 'Conservation for the People', *Scientific American* 297.4: 50–57.

Kashwan, Prakash. 2017. *Democracy in the Woods: Environmental Conservation and Social Justice in India, Tanzania, and Mexico* (Oxford: Oxford University Press).

Kill, Jutta and Giulia Franchi. 2016. *Rio Tinto's Biodiversity Offset in Madagascar: Double Landgrab in the Name of Biodiversity?* (Montevideo: World Rainforest Movement).

Kim, Ki-Hyun, Ehsanul Kabir, and Shamin Kabir. 2015. 'A Review on the Human Health Impact of Airborne Particulate Matter', *Environment International* 74: 136–43.

Kinzig, Ann et al. 2011. 'Paying for Ecosystem Services—Promise and Peril', *Science* 334.6056: 603–4.

Klosko, George. 1987. 'Presumptive Benefit, Fairness, and Political Obligation', *Philosophy & Public Affairs* 16.3: 241–59.

Koenigstein, Stefan et al. 2016. 'Modelling Climate Change Impacts on Marine Fish Populations: Process-based Integration of Ocean Warming, Acidification and Other Environmental Drivers', *Fish and Fisheries* 17.4: 972–1004.

Koh, Niak Sian et al. 2019. 'How Much of a Market Is Involved in a Biodiversity Offset? A Typology of Biodiversity Offset Policies', *Journal of Environmental Management* 232: 679–91.

Kolbert, Elizabeth. 2014. *The Sixth Great Extinction* (London: Bloomsbury).

Kopnina, Helen. 2016. 'Half the Earth for People (or more)? Addressing Ethical Questions in Conservation', *Biological Conservation* 203: 176–85.

Kopnina, Helen et al. 2018. 'The "Future of Conservation" Debate: Defending Ecocentrism and the Nature Needs Half Movement', *Biological Conservation* 217: 140–48.

Korsgaard, Christine. 2018. *Fellow Creatures: Our Obligations to the Other Animals* (Oxford: Oxford University Press).

Kothari, Ashish. 2021. 'Half-Earth or Whole-Earth? Green or Transformative Recovery? Where Are the Voices from the Global South?' *Oryx* 55.2: 161–62.

Kotsakis, Andreas. 2021. *The Use of Biodiversity in International Law* (Abingdon: Routledge).

Kühl, Hjalmar et al. 2017. 'The Critically Endangered Western Chimpanzee Declines by 80%', *American Journal of Primatology* 79.9: e22681.

Ladwig, Bernd. 2015. 'Against Wild Animal Sovereignty: An Interest-based Critique of Zoopolis', *Journal of Political Philosophy* 23.3: 282–301.

Lawler, Odette et al. 2021. 'The COVID-19 Pandemic Is Intricately Linked to Biodiversity Loss and Ecosystem Health', *The Lancet Planetary Health* 5.11: e840–50.

Leach, Melissa and Ian Scoones. 2013. 'Carbon Forestry in West Africa: The Politics of Models, Measures and Verification Processes', *Global Environmental Change* 23.5: 957–67.

Leadley, Paul et al. 2022. 'Achieving Global Biodiversity Goals by 2050 Requires Urgent and Integrated Actions', *One Earth* 5: 597–603.

Leclere, David et al. 2020. 'Bending the Curve of Terrestrial Biodiversity Needs an Integrated Strategy', *Nature* 585: 551–56.

Legagneux, Pierre et al. 2018. 'Our House Is Burning: Discrepancy in Climate Change vs. Biodiversity Coverage in the Media as Compared to Scientific Literature', *Frontiers in Ecology and Evolution* 5: 175.

Lennox, Gareth and Paul Armsworth. 2013. 'The Ability of Landowners and their Cooperatives to Leverage Payments Greater than Opportunity Costs from Conservation Contracts', *Conservation Biology* 27.3: 625–34.

Lenzen, Manfred et al. 2012. 'International Trade Drives Biodiversity Threats in Developing Nations', *Nature* 486.7401: 109–12.

Leong, Misha et al. 2018. 'Biodiversity and Socioeconomics in the City: A Review of the Luxury Effect', *Biology Letters* 14: 20180082.

Lepora, Chiara and Robert Goodin. 2013. *On Compromise and Complicity* (Oxford: Oxford University Press).

Lin, David et al. 2018. 'Ecological Footprint Accounting for Countries: Updates and Results of the National Footprint Accounts, 2012–2018', *Resources* 7.3: 58.

Lindhjem, Henrik and Yohei Mitani. 2012. 'Forest Owners' Willingness to Accept Compensation for Voluntary Conservation: A Contingent Valuation Approach', *Journal of Forest Economics* 18.4: 290–302.

Locke, Harvey. 2014. 'Nature Needs Half: A Necessary and Hopeful New Agenda for Protected Areas in North America and around the World', *The George Wright Forum* 31.3: 359–71.

Lundquist, Carolyn. 2021. 'Transformative Scenarios for Biodiversity Conservation and Sustainability', *Conservation Letters* 14.2: e12772.

Maas, Bea et al. 2021. 'Women and Global South Strikingly Underrepresented Among Top-publishing Ecologists', *Conservation Letters* 14.4: e12797.

Machovina, Brian, Kenneth Feeley, and William Ripple. 2015. 'Biodiversity Conservation: The Key is Reducing Meat Consumption', *Science of the Total Environment* 536: 419–31.

Maier, Donald. 2013. *What's So Good about Biodiversity?* (Dordrecht: Springer).

Mancilla, Alejandra. 2022. 'From Sovereignty to Guardianship in Ecoregions', *Journal of Applied Philosophy* 40.4: 608–23.

Maron, Martine et al. 2015a. 'Stop Misuse of Biodiversity Offsets', *Nature* 523: 401–3.

Maron, Martine et al. 2015b. 'Locking in Loss: Baselines of Decline in Australian Biodiversity Offset Policies', *Biological Conservation* 192: 504–12.

Maron, Martine et al. 2018. 'The Many Meanings of No Net Loss in Environmental Policy', *Nature Sustainability* 1.1: 19–27.

Maron, Martine et al. 2020. 'Global No Net Loss of Natural Ecosystems', *Nature Ecology & Evolution* 4.1: 46–49.

Marris, Emma. 2011. *Rambunctious Garden: Saving Nature in a Post-wild World* (London: Bloomsbury).

Marshall, Erica et al. 2020. 'What Are We Measuring? A Review of Metrics Used to Describe Biodiversity in Offsets Exchanges', *Biological Conservation* 241: 108250.

Martin, Adrian. 2017. *Just Conservation: Biodiversity, Wellbeing and Sustainability* (Abingdon: Routledge).

Martin, Adrian, Anne Akol, and Nicole Gross-Camp. 2015. 'Towards an Explicit Justice Framing of the Social Impacts of Conservation', *Conservation and Society* 13.2: 166–78.

Marvier, Michelle. 2014. 'New Conservation is True Conservation', *Conservation Biology* 28.1: 1–3.

Marvier, Michelle, Peter Kareiva, and Robert Lalasz. 2011. 'Conservation in the Anthropocene: Beyond Solitude and Fragility', *Breakthrough Journal* 2 [online] https://thebreakthrough.org/journal/issue-2/conservation-in-the-anthropocene

Massarella, Kate et al. 2021. 'Transformation Beyond Conservation: How Critical Social Science Can Contribute to a Radical New Agenda in Biodiversity Conservation', *Current Opinion in Environmental Sustainability* 49: 79–87.

Maxwell, Sean et al. 2010. 'Biodiversity: The Ravages of Guns, Nets and Bulldozers', *Nature News* 536.7615: 143–45.

Mayer, Robert. 2007. 'What's Wrong with Exploitation?' *Journal of Applied Philosophy* 24.2: 137–50.

McCarthy, Donal et al. 2012. 'Financial Costs of Meeting Global Biodiversity Conservation Targets: Current Spending and Unmet Needs', *Science* 338.6109: 946–49.

McCormick, Rachel. 2017. 'Does Access to Green Space Impact the Mental Well-being of Children: A Systematic Review', *Journal of Pediatric Nursing* 37: 3–7.

McFarland, Will and Shelagh Whitley. 2015. *Subsidies to Key Commodities Driving Forest Loss* (London: Overseas Development Institute).

McKee, Jeffrey et al. 2004. 'Forecasting Global Biodiversity Threats Associated with Human Population Growth', *Biological Conservation* 115.1: 161–64.

McKinnon, Catriona. 2012. *Climate Change and Future Justice: Precaution, Compensation and Triage* (Abingdon: Routledge).

McLaren, Duncan and Louise Carver. 2023. 'Disentangling the Net from the Offset: Learning for Net-Zero Climate Policy from an Analysis of 'No Net Loss' in Biodiversity', *Frontiers in Climate* 5: 1197608.

McShane, Katie. 2007. 'Why Environmentalists Shouldn't Give Up on Intrinsic Value', *Environmental Ethics* 29.1: 43–61.

McShane, Katie. 2017. 'Is Biodiversity Intrinsically Valuable? (And What Might That Mean?)', in Justin Garson, Anya Plutynski, and Sahotra Sarkar (eds) *Routledge Handbook of Philosophy of Biodiversity* (London: Routledge), 155–67.

Meadows, Donella et al. 1972. *The Limits to Growth: A Report for the Club of Rome* (New York: Universe Books).

Mehrabi, Zia, Erle Ellis, and Navin Ramankutty. 2018. 'The Challenge of Feeding the World While Conserving Half the Planet', *Nature Sustainability* 1: 409–12.

Mikkelson, Gregory. 2013. 'Growth Is the Problem; Equality Is the Solution', *Sustainability* 5: 432–39.

Mikkelson, Gregory, Andrew Gonzalez, and Garry Peterson. 2009. 'Economic Inequality Predicts Biodiversity Loss', *Conservation Biology* 23: 1304–13.

Milburn, Josh and Sara Van Goozen. 2020. 'Counting Animals in War: First Steps Towards an Inclusive Just-war Theory', *Social Theory and Practice* 47.4: 657–85.

Miller, David. 2007. *National Responsibility and Global Justice* (Oxford: Oxford University Press).

Miller, David. 2009. Global Justice and Climate Change: How Should Responsibilities Be Distributed? *Tanner Lectures on Human Values* 28: 117–56.

Miller, Brian, Micheal Soulé, and John Terborgh. 2014. 'New Conservation—Or Surrender to Development', *Animal Conservation* 17.6: 509–15.

Mills, Julianne and Thomas Waite. 2009. 'Economic Prosperity, Biodiversity Conservation, and the Environmental Kuznets Curve', *Ecological Economics* 68.7: 2087–95.

Mirza, M. Usman et al. 2020. 'Institutions and Inequality Interplay Shapes the Impact of Economic Growth on Biodiversity Loss', *Ecology and Society* 25.4: 39–49.

Moellendorf, Darrel. 2009. *Global Inequality Matters* (Basingstoke: Palgrave Macmillan).

Moellendorf, Darrel. 2014. *The Moral Challenge of Dangerous Climate Change* (Cambridge: Cambridge University Press).

Moellendorf, Darrel. 2022. *Mobilizing Hope: Climate Change and Global Poverty* (Oxford: Oxford University Press).

Molnar, Jennifer et al. 2008. 'Assessing the Global Threat of Invasive Species to Marine Biodiversity', *Frontiers in Ecology and the Environment* 6.9: 485–92.

Montoya, Robert. 2022. *Power of Position: Classification and the Biodiversity Sciences* (Cambridge, MA: MIT Press).

Moore, Jason. 2017. 'The Capitalocene, Part I: On the Nature and Origins of our Ecological Crisis', *The Journal of Peasant Studies* 44.3: 594–630.

Moreno-Mateos, David et al. 2015. 'The True Loss Caused by Biodiversity Offsets', *Biological Conservation* 192: 552–59.

Muka, Samantha and Chris Zarpentine. 2023. 'Southern Resident Orca Conservation: Practical, Ethical and Political Issues', to be published in *Ethics, Policy & Environment* [preprint] https://doi.org/10.1080/21550085.2023.2175453

Murdock, Esme. 2021. 'Conserving Dispossession? A Genealogical Account of the Colonial Roots of Western Conservation', *Ethics, Policy & Environment* 24.3: 235–49.

Naeem, Shahid, J. Emmett Duffy, and Erika Zavaleta. 2012. 'The Functions of Biological Diversity in an Age of Extinction', *Science* 336.6087: 1401–6.

Naidoo, Robin et al. 2006. 'Integrating Economic Costs into Conservation Planning', *Trends in Ecology & Evolution* 21.12: 681–7.

Naidoo, Robin and Wiktor Adamowicz. 2006. 'Modeling Opportunity Costs of Conservation in Transitional Landscapes', *Conservation Biology* 20.2: 490–500.

Napoletano, Brian and Brett Clark. 2020. 'An Ecological-Marxist Response to the Half-Earth Project', *Conservation and Society* 18.1: 37–49.

Neuteleers, Stijn. 2022. 'Survey Article: Trading Nature: When Are Environmental Markets (Un)desirable?' *Journal of Political Philosophy* 30.1: 116–39.

Newman, Jonathan, Gary Varner, and Stefan Linquist (2017) *Defending Biodiversity* (Cambridge: Cambridge University Press).

Niemeyer, Simon. 2014. 'A Defence of (Deliberative) Democracy in the Anthropocene', *Ethical Perspectives* 21: 15–45.

Neumann, Roderick. 1998. *Imposing Wilderness: Struggles Over Livelihood and Nature Preservation in Africa* (Berkeley: University of California Press).

Newbold, Tim et al. 2016. 'Has Land Use Pushed Terrestrial Biodiversity Beyond the Planetary Boundary? A Global Assessment', *Science* 353.6296: 288–91.

Norse, Elliott et al. 1986. *Conserving Biological Diversity in Our National Forests* (Washington D.C.: The Wilderness Society).

Noss, Reed et al. 2012. 'Bolder Thinking for Conservation', *Conservation Biology* 26: 1–4.

Nowak, Katarzyna. 2016. *CITES Alone Cannot Combat Illegal Wildlife Trade*. Policy Insight 34. (Johannesburg: South African Institute of International Affairs)

Nussbaum, Martha. 2006. *Frontiers of Justice: Disability, Nationality, Species Membership*. (Cambridge, MA: Belknap Press).

Nussbaum, Martha. 2022. *Justice for Animals: Our Collective Responsibility* (New York: Simon and Schuster).

Ocampo-Ariza, Carolina et al. 2023. 'Global South leadership towards Inclusive Tropical Ecology and Conservation', *Perspectives in Ecology and Conservation* 21.1: 17–24.

Ogar, Edwin et al. 2020. 'Science Must Embrace Traditional and Indigenous Knowledge to Solve Our Biodiversity Crisis', *One Earth* 3.2: 162–65.

Oldekop, Johan et al. 2016. 'A Global Assessment of the Social and Conservation Outcomes of Protected Areas', *Conservation Biology* 30.1: 133–41.

O'Neill, John. 1992. 'The Varieties of Intrinsic Value', *The Monist* 75.2: 119–37.

Österblom, Henrik et al. 2015. 'Transnational Corporations as "Keystone Actors" in Marine Ecosystems', *PLoS One* 10.5: e0127533.

Pandit, Ram and David Laband. 2009. 'Economic Well-being, the Distribution of Income and Species Imperilment', *Biodiversity Conservation* 18: 3219–33.

Pascual, Unai et al. 2021. 'Biodiversity and the Challenge of Pluralism', *Nature Sustainability* 4: 567–72.

Passmore, John. 1974. *Man's Responsibility for Nature* (London: Duckworth).

Peluso, Nancy L. 1993. 'Coercing Conservation? The Politics of State Resource Control', *Global Environmental Change* 3.2: 199–217.

Perelman, Marcus. 2007. 'Primitive Accumulation from Feudalism to Neoliberalism', *Capitalism Nature Socialism* 18.2: 44–61.

Pettit, Philip. 1997. *Republicanism: A Theory of Freedom and Government* (Oxford: Oxford University Press).

Phalan, Ben et al. 2018. 'Avoiding Impacts on Biodiversity through Strengthening the First Stage of the Mitigation Hierarchy', *Oryx* 52: 316–24.

Pickering, Jonathan et al. 2022. 'Rethinking and Upholding Justice and Equity in Transformative Biodiversity Governance', in Ingrid Visseren-Hamakers and Marcel Kok (eds) *Transforming Biodiversity Governance* (Cambridge: Cambridge University Press), 155–78.

Pimm, Stuart, Clinton Jenkins, and Binbin Li. 2018. 'How to Protect Half of Earth to Ensure It Protects Sufficient Biodiversity', *Science Advances* 4.8: eaat2616.

Placani, Adriana and Stearns Broadhead. 2022. 'Moral Dimensions of Offsetting Luxury Emissions', *Ethics, Policy & Environment* 77: 1–19.

Pogge, Thomas. 2002. *World Poverty and Human Rights* (Cambridge: Polity).

Porter-Bolland, Luciana et al. 2012. 'Community Managed Forests and Forest Protected Areas: An Assessment of their Conservation Effectiveness Across the Tropics', *Forest Ecology and Management* 268: 6–17.

Poudyal, Mahesh et al. 2018a. 'Who Bears the Cost of Forest Conservation?' *PeerJ* 6: e5106.

Poudyal, Mahesh et al. 2018b. 'Household Economy, Forest Dependency & Opportunity Costs of Conservation in Eastern Rainforests of Madagascar', *Nature: Scientific Data* 5: 180225.

Prendergast, Drew and Troy Vettese. 2021. 'The Humanization of Nature and Half-Earth Socialism', *International Labor and Working-Class History* 99.1: 15–23.

Purvis, Andy et al. 2000. 'Predicting Extinction Risk in Declining Species', *Proceedings of the Royal Society: Biological Sciences* 267.1456: 1947–52.

Raja, Nussaïbah et al. 2022. 'Colonial History and Global Economics Distort Our Understanding of Deep-time Biodiversity', *Nature Ecology & Evolution* 6.2: 145–54.

Rands, Michael et al. 2010. 'Biodiversity Conservation: Challenges Beyond 2010', *Science* 329.5997: 1298–303.

Rawls, John. 1971. *A Theory of Justice* (Cambridge, MA: Harvard University Press).

Rawls, John. 1999. *The Law of Peoples* (Cambridge, MA: Harvard University Press).

Rockström, Johan et al. 2009. 'A Safe Operating Space for Humanity', *Nature* 461.7263: 472–5.

Rohwer, Yasha and Emma Marris. 2015. 'Is There a Prima Facie Duty to Preserve Genetic Integrity in Conservation Biology?' *Ethics, Policy & Environment* 18.3: 233–47.

Rolston, Holmes. 1985. 'Duties to Endangered Species', *BioScience* 35.11: 718–26.

Rolston, Holmes. 2012. *A New Environmental Ethics: The Next Millennium for Life on Earth* (London: Routledge).

Rowlands, Mark. 2021. *World on Fire: Humans, Animals, and the Future of the Planet* (Oxford: Oxford University Press).

Salzman, James et al. 2018. 'The Global Status and Trends of Payments for Ecosystem Services', *Nature Sustainability* 1.3: 136–44.

Samuelsson, Lars. 2010. 'Reasons and Values in Environmental Ethics', *Environmental Values* 19: 517–35.

Sanderson, Steven and Kent Redford. 2003. 'Contested Relationships between Biodiversity Conservation and Poverty Alleviation', *Oryx* 37.4: 389–90.

Santana, Carlos. 2014. 'Save the Planet: Eliminate Biodiversity', *Biology & Philosophy* 29.6: 761–80.

Sarkar, Sahotra. 1999. 'Wilderness Preservation and Biodiversity Conservation—Keeping Divergent Goals Distinct', *BioScience* 49.5: 405–12.

Sarkar, Sahotra. 2006. 'Ecological Diversity and Biodiversity as Concepts for Conservation Planning', *Acta Biotheoretica* 54: 133–40.

Satz, Debra. 2010. *Why Some Things Should Not Be for Sale: The Moral Limits of Markets* (Oxford: Oxford University Press).

Schaffartzik, Anke, Juan Antonio Duro, and Fridolin Krausmann. 2019. 'Global Appropriation of Resources Causes High International Material Inequality—Growth Is Not the Solution', *Ecological Economics* 163: 9–19.

Scheidel, Arnim et al. 2020. 'Environmental Conflicts and Defenders: A Global Overview', *Global Environmental Change* 63: 102104.

Schell, Christopher et al. 2020. 'The Ecological and Evolutionary Consequences of Systemic Racism in Urban Environments', *Science* 369.6510: eaay4497.

Schleicher, Judith et al. 2017. 'Conservation Performance of Different Conservation Governance Regimes in the Peruvian Amazon', *Scientific Reports* 7.1: 1–10.

Schleicher, Judith et al. 2019. 'Protecting Half of the Planet Could Directly Affect over One Billion People', *Nature Sustainability* 2.12: 1094–6.

Schlosberg, David. 2007. *Defining Environmental Justice: Theories, Movements, and Nature* (Oxford: Oxford University Press).

Schuster, Richard et al. 2019. 'Vertebrate Biodiversity on Indigenous-managed Lands in Australia, Brazil, and Canada Equals that in Protected Areas', *Environmental Science & Policy* 101: 1–6.

Sebo, Jeff. 2021. 'How to Count Animals, More or Less (book review)', *Mind* 130: 518: 689–97.

Sebo, Jeff. 2022. *Saving Animals, Saving Ourselves* (Oxford: Oxford University Press).

Sepúlveda, Julio and Sebastian Cantarero. 2022. 'Phytoplankton Response to a Warming Ocean', *Science* 376.6600: 1378–79.

Shandra, John et al. 2009. 'Ecologically Unequal Exchange, World Polity, and Biodiversity Loss', *International Journal of Comparative Sociology* 50.3–4: 285–310.

Shandra, John et al. 2010. 'Debt, Structural Adjustment, and Biodiversity Loss: A Cross-National Analysis of Threatened Mammals and Birds', *Human Ecology Review* 17.1: 18–33.

Shue, Henry. 1980. *Basic Rights: Subsistence, Affluence, and US Foreign Policy* (Princeton: Princeton University Press).

Shue, Henry. 2014. *Climate Justice: Vulnerability and Protection* (Oxford: Oxford University Press).

Simpson, Katherine et al. 2021. 'Incentivising Biodiversity Net Gain with An Offset Market', *Q Open* 1.1: qoab004.

Sims, Katherine and Jennifer Alix-Garcia. 2017. 'Parks versus PES: Evaluating Direct and Incentive-based Land Conservation in Mexico', *Journal of Environmental Economics and Management* 86: 8–28.

Singer, Peter. 1972. 'Famine, Affluence and Morality', *Philosophy & Public Affairs* 1.3: 229–43.

Singer, Peter. 1974. 'All Animals are Equal', *Philosophic Exchange*, 1.5:103–16.

Singer, Peter. 1979. 'Killing Humans and Killing Animals', *Inquiry* 22(1_4): 145–56.

Siurua, Hanna. 2006. 'Nature above People: Conservation in the South', *Ethics & The Environment* 11.1: 71–96.

Sonter, Laura et al. 2018. 'Biodiversity Offsets May Miss Opportunities to Mitigate Impacts on Ecosystem Services', *Frontiers in Ecology and the Environment* 16.3: 143–8.

Soto-Navarro, Carolina A. et al. 2021. 'Toward a Multidimensional Biodiversity Index for National Application', *Nature Sustainability* 4: 933–42.

Soulé, Michael. 1985. 'What Is Conservation Biology?' *BioScience* 35.11: 727–34.

Stark, Keila et al. 2021. 'Importance of Equitable Cost Sharing in the Convention on Biological Diversity's Protected Area Agenda', *Conservation Biology* 36.2: 1–4.

Steffen, Will et al. 2015. 'Planetary Boundaries: Guiding Human Development on a Changing Planet', *Science* 347.6223: 1259855.

Sterelny, Kim. 2021. *The Pleistocene Social Contract: Culture and Cooperation in Human Evolution* (Oxford: Oxford University Press).

Stern, David. 2004. 'The Rise and Fall of the Environmental Kuznets Curve', *World Development* 32.8: 1419–39.

Stilz, Anna. 2017. 'Settlement, Expulsion, and Return', *Politics, Philosophy & Economics* 16.4: 351–74.

Sumaila, U. Rashid et al. 2019. 'Updated Estimates and Analysis of Global Fisheries Subsidies', *Marine Policy* 109: 103695.

Swartz, Wilf et al. 2010. 'Sourcing Seafood for the Three Major Markets: The EU, Japan and the USA', *Marine Policy* 34: 1366–73.

Szende, Jennifer. 2022. 'Relational Value, Land, and Climate Justice', *Journal of Global Ethics* 18.1: 118–33.

Tadesse, Tewodros et al. 2021. 'Willingness to Accept Compensation for Afromontane Forest Ecosystems Conservation', *Land Use Policy* 105: 105382.

Tan, Kok-Chor. 2004. *Justice Without Borders* (Cambridge: Cambridge University Press).

Tan, Kok-Chor. 2021. 'Just Conservation: The Question of Justice in Global Wildlife Conservation', *Philosophy Compass* 16.2: 1–12.

Tan, Kok-Chor. 2023. 'Climate Reparations: Why the Polluter Pays Principle is neither Unfair nor Unreasonable', *WIRES Climate Change* 14.4: e827.

Tauli-Corpuz, Vicky, et al. 2020. 'Cornered by PAs: Adopting Rights-based Approaches to Enable Cost-effective Conservation and Climate Action', *World Development* 130: 104923.

Terborgh, John. 1999. *Requiem for Nature* (Washington, D.C: Island Press).

Thondhlana, Gladman et al. 2020. 'Non-material Costs of Wildlife Conservation to Local People and Their Implications for Conservation Interventions', *Biological Conservation* 246: 108578.

Thorn, Sian et al. 2018. 'Effectiveness of Biodiversity Offsets: An Assessment of a Controversial Offset in Perth, Western Australia', *Biological Conservation* 228: 291–300.

Timmer, Vanessa and Calestous Juma. 2005. 'Taking Root: Biodiversity Conservation and Poverty Reduction Come together in the Tropics', *Environment: Science and Policy for Sustainable Development* 47.4: 24–44.

Tuck, Eve and K. Wayne Yang. 2012. 'Decolonization Is Not a Metaphor', *Decolonization: Indigeneity, Education and Society* 1.1: 1–40.

United Nations Department of Economic and Social Affairs. 2019. *World Population Prospects 2019: Highlights* (New York: United Nations).

Vanderheiden, Steve. 2008. *Atmospheric Justice: A Political Theory of Climate Change* (Oxford: Oxford University Press).

Van der Sluijs, Jeroen and Nora Vaage. 2016. 'Pollinators and Global Food Security: The Need for Holistic Global Stewardship', *Food Ethics* 1.1: 75–91.

Venter, Oscar et al. 2016. 'Sixteen Years of Change in the Global Terrestrial Human Footprint and Implications for Biodiversity Conservation', *Nature Communications* 7.1: 1–11.

Vettese, Troy. 2020. 'A Marxist Theory of Extinction', *Salvage* 7 [online], 25 May. https://salvage.zone/a-marxist-theory-of-extinction/

Vogel, Steven. 2011. 'Why "Nature" Has No Place in Environmental Philosophy', in Gregory Kaebnick (ed) *The Ideal of Nature: Debates About Biotechnology and the Environment* (Baltimore, MA: Johns Hopkins University Press), 84–97.

Voytenko, Yuliya et al. 2016. 'Urban Living Labs for Sustainability and Low Carbon Cities in Europe: Towards a Research Agenda', *Journal of Cleaner Production* 123: 45–54.

Vucetich, John et al. 2018. 'Just Conservation: What Is It and Should We Pursue It?' *Biological Conservation* 221: 23–33.

Waldron, Anthony et al 2017. 'Reductions in Global Biodiversity Loss Predicted from Conservation Spending', *Nature* 551: 364–67.

Waldron, Anthony et al. 2020. Protecting 30% of the Planet for Nature: Costs, Benefits and Economic Implications. Campaign for Nature Working Paper (Cambridge: Conservation Research Institute).

Walker, Susan et al. 2009. 'Why Bartering Biodiversity Fails', *Conservation Letters* 2: 149–57.

Wall, Steven and David Sobel. 2021. 'A Robust Hybrid Theory of Well-being', *Philosophical Studies* 178.9: 2829–51.

Watson, James et al. 2018a. 'Protect the Last of the Wild', *Nature* 563: 27–30.

Watson, James et al. 2018b. 'The Exceptional Value of Intact Forest Ecosystems', *Nature Ecology & Evolution* 2.4: 599–610.

Watson, James et al. 2021. 'Talk is Cheap: Nations Must Act Now to Achieve Long-term Ambitions for Biodiversity', *One Earth* 4: 897–900.

Wauchope, Hannah, Justine Shaw, and Aleks Terauds. 2019. 'A Snapshot of Biodiversity Protection in Antarctica', *Nature Communications* 10.1: 1–6.

Weatherley-Singh, Janice et al. 2022. 'Transformative Biodiversity Governance for Protected and Conserved Areas', in Ingrid Visseren-Hamakers and Marcel Kok (eds) *Transforming Biodiversity Governance* (Cambridge: Cambridge University Press), 221–43.

Wenar, Leif. 2015. *Blood Oil* (Oxford: Oxford University Press).

World Health Organization. 2015. *Connecting Global Priorities: Biodiversity and Human Health* (Geneva: World Health Organization).

Whyte, Kyle. 2018. 'Settler Colonialism, Ecology, and Environmental Injustice', *Environment and Society: Advances in Research* 9: 125–44.

Wiedmann, Thomas et al. 2015. 'The Material Footprint of Nations', *Proceedings of the National Academy of Sciences of the United States of America* 112.20: 6271–6.

Wilhere, George. 2021. 'A Paris-like Agreement for Biodiversity Needs IPCC-like Science', *Global Ecology and Conservation* 28: e01617.

Wilson, Edward. 2003. *The Future of Life* (New York: Random House).

Wilson, Edward. 2016. *Half-Earth: Our Planet's Fight for Life* (New York: W.W. Norton).

Wilson, Kerrie, Jacqueline England, and Shaun Cunningham. 2017. 'Biodiversity Indicators Need to be Fit for Purpose', in Justin Garson, Anya Plutynski, and Sahotra Sarkar (eds) *Routledge Handbook of Philosophy of Biodiversity* (London: Routledge), 229–40.

Wolch, Jennifer, Jason Byrne, and Joshua Newell. 2014. 'Urban Green Space, Public Health, and Environmental Justice: The Challenge of Making Cities "Just Green Enough"', *Landscape and Urban Planning* 125: 234–44.

Woodhouse, Emily et al. 2021. 'Rethinking Entrenched Narratives About Protected Areas and Human Well-being in the Global South', *UCL Open: Environment* 4: 1–32. https://doi.org/ 10.14324/111.444/ucloe.000050

World Economic Forum. 2022. *Biodiversity Credits: Unlocking Financial Markets for Nature-positive Outcomes* (Geneva: World Economic Forum).

Wulf, Andrea. 2015. *The Invention of Nature: The Adventures of Alexander von Humboldt, the Lost Hero of Science* (London: Hachette).

Wündisch, Joachim. 2017. 'Does Excusable Ignorance Absolve of Liability for Costs?' *Philosophical Studies* 174: 837–51.

Worldwide Fund for Nature (WWF). 2022. *Living Planet Report 2022* (Gland, Switzerland: WWF).

Wyborn, Carina et al. 2020a. 'An Agenda for Research and Action towards Diverse and Just Futures for Life on Earth', *Conservation Biology* 35.4: 1086–97.

Wyborn, Carina et al. 2020b. 'Imagining Transformative Biodiversity Futures', *Nature Sustainability* 3.9: 670–2.

Yankah, Ekow. 2019. 'Whose Burden to Bear? Privilege, Lawbreaking and Race', *Criminal Law and Philosophy* 16: 13–28.

Youatt, Rafi. 2015. *Counting Species: Biodiversity in Global Environmental Politics* (Minneapolis, MN: University of Minnesota Press).

Zaccai, Edwin and William Adams. 2012. 'How Far Are Biodiversity Loss and Climate Change Similar as Policy Issues?' *Environment, Development and Sustainability* 14: 557–71.

Zaitchik, Alexander. 2018. 'How Conservation Became Colonialism', *Foreign Policy* 229: 56–63.

zu Ermgassen, Sophus et al. 2019a. 'The Ecological Outcomes of Biodiversity Offsets under "No Net Loss"', *Policies: A Global Review* 12.6: e12664.

zu Ermgassen, Sophus et al. 2019b. 'The Role of "No Net Loss" Policies in Conserving Biodiversity Threatened by the Global Infrastructure Boom', *One Earth* 1: 305–15.

zu Ermgassen, Sophus et al. 2020. 'The Hidden Biodiversity Risks of Increasing Flexibility in Biodiversity Offset Trades', *Biological Conservation* 252: 108861

Index